Dam Break Modelling, Risk Assessment and Uncertainty Analysis for Flood Mitigation

Dam Break Modelling, Risk Assessment and Uncertainty Analysis for Flood Mitigation

DISSERTATION

Submitted in fulfillment of the requirements of
the Board for Doctorates of Delft University of Technology
and of the Academic Board of the UNESCO-IHE Institute for Water Education
for the Degree of DOCTOR
to be defended in public
on 24 September 2007 at 10:00 hours
in Delft, the Netherlands

by

Migena ZAGONJOLLI

born in Tirana, Albania
Bachelor of Science in Hydraulic Engineering, Polytechnic University of Tirana
Master of Science in Hydroinformatics, UNESCO-IHE

CRC Press
Taylor & Francis Group
Boca Raton London New York

CRC Press is an imprint of the
Taylor & Francis Group, an **informa** business

This dissertation has been approved by the promotor: Prof.dr.ir. A.E. Mynett

Members of the Awarding Committee:

Chairman	Rector Magnificus TU Delft, the Netherlands
Vice Chairman	Rector UNESCO–IHE, the Netherlands
Prof.dr.ir. A.E. Mynett	TU Delft / UNESCO-IHE, the Netherlands, Promotor
Prof.dr.ir. H.J. de Vriend	TU Delft / WL | Delft Hydraulics, the Netherlands
Prof.drs.ir. J.K. Vrijling	TU Delft, the Netherlands
Prof.dr. N. Wright	UNESCO–IHE / TU Delft, the Netherlands
Prof.dr. Y. Zech	Université Catholique de Louvain, Belgium
Prof.dr. K. Takeuchi	ICHARM Centre, PWRI, Japan
Prof.dr.ir. G. S. Stelling	TU Delft, the Netherlands, Reserve Member

CRC Press
Taylor & Francis Group
6000 Broken Sound Parkway NW, Suite 300
Boca Raton, FL 33487-2742

First issued in hardback 2018

© 2007 by Taylor and Francis Group, LLC
CRC Press is an imprint of Taylor & Francis Group, an Informa business

No claim to original U.S. Government works

ISBN-13: 978-0-415-45594-7 (pbk)
ISBN-13: 978-1-138-46581-7 (hbk)

Visit the Taylor & Francis Web site at
http://www.taylorandfrancis.com

and the CRC Press Web site at
http://www.crcpress.com

This thesis is dedicated to my parents, for their endless love, encouragement and support, and for always being near me even when I was far away.

Summary

More than 800,000 dams and thousands of kilometers of dikes have been constructed around the world. However, the history of the construction of dams and dikes (in this thesis referred to as 'structures') coexists with the history of their collapse. Hundreds of dam failure events were reported the past centuries, but still today dikes breach every year due to high water levels, often with catastrophic consequences. In the Netherlands the storm surge of 1953 with 1850 casualties led to the construction of the famous Delta Works. One more recent example is the breaching of the New Orleans levee systems during hurricane Katrina in August 2005, which caused prolonged flooding with 1,300 casualties as well as tens of billions of dollars of economic and social damage.

Effects of climate change are likely to cause more severe flow conditions within the life span of existing structures, leading to increased safety concerns. If a structure fails, the release of (large quantities of) water may threaten the lives of people as well as property in the downstream areas. Likely loss of life depends on actual water depth and flow velocity, the geographical distribution of the population, warning time necessary to reach them and their awareness at the time of disaster. Warning messages released in advance can be an important factor for saving lives. Hence, developing and improving risk assessment and flood mitigation models is becoming increasingly important and can be considered a necessity to reduce human casualties and economic damage.

The aim of this research is to develop a framework and explore techniques for modelling dam and dike failure events, as well as to develop novel approaches for risk assessment. Numerical, statistical and constraint based methods are applied to breach modelling and flood water mitigation. A new breach model (BREADA) is developed for simulating the gradual failure of a structure due to overtopping which is validated against historical dam failure events. In order to explore the accuracy of breach models in a different way, and to try to extract possible additional information from available data of recorded dam failure events, we use data mining techniques that have been successfully applied in the field of hydroinformatics.

While physically based methods require proper understanding of all processes occurring during dam breaching, data mining methods rely only on recorded data of dam failure events. In this research we apply data mining techniques including Artificial Neural Networks and Instance Based Learning for predicting dam breach characteristics and peak outflows. Despite the shortage of documented data, this research demonstrates there is a possibility to improve the presently available empirical relations and prediction capabilities of physically based models by complementing with data mining techniques.

The analysis of a potential failure event for existing structures (especially for large dams) is essential for planning and organizing emergency procedures that anticipate and mitigate downstream damages in case of disaster. In this thesis we analyse a hypothetical failure event for the Bovilla Dam near Tirana, Albania, and explore potential mitigation measures in case of a worst case scenario. A comparison is made between the results of the developed BREADA model, other available breach modelling formulations, and empirical techniques in order to get a range of peak outflows. Flood routing is carried out using WL | Delft Hydraulics' hydrodynamic modelling package Sobek 1D2D. Sensitivity analysis is carried out to identify uncertainties associated with dam failure analysis. The model is used to identify areas prone to flooding, to assess the risk involved, as well as to take measures to reduce flood damage and develop emergency plans.

Clearly, structural failure events pose a significant threat not only to human life but also to the environment and in general also to economic development. With such catastrophic consequences in mind, it is essential to investigate not only mechanisms for predicting these failure events but also to reduce their risk of occurrence. Traditional approaches focus on establishing very small probabilities of occurrence of extreme events. However, if such event were to occur - despite its low probability value - the consequences may be very severe, like in the case of New Orleans.

In this research, instead of focusing on methodologies to minimise the failure probability of a structure, we consider an alternative approach that aims at decreasing the consequences of a flood event. A numerical–constraint based model is developed for evaluating risk and mitigating consequences in a system of polders or low–lying areas. The model is capable of simultaneously evaluating different flood mitigation scenarios in a very short time by utilizing algorithms based on 'graph theory'. The results of a case study which takes into account different objective functions such as storage capacities and economical values of a multiply connected polder system, look quite promising for flood risk mitigation.

The approach developed in this thesis can be used to complement existing practices of flood modelling, which are traditionally carried out by simulating the consequences of a forecasted or assumed flood event and elaborated into a few typical 'what - if' scenarios. The 'lightweight' numerical–constraint based technique proposed in this thesis is capable of evaluating many scenarios in a very short period of time by first determining the most feasible scenarios, which can then be modelled in more detail using a conventional hydrodynamic simulation approach. In this way computation time is considerably reduced while focusing on the most feasible options is ensured. Clearly a combination of these two methods can either be achieved by enhancing a hydrodynamic modelling package with an optional numerical–constraint based approach, or vice versa.

Samenvatting

Wereldwijd zijn meer dan 800,000 dammen en vele duizenden kilometers dijk aangeled. De geschiedenis van dammen en dijken gaat echter hand in hand met de geschiedenis van hun falen. Honderden gevallen van het bezwijken van dijken hebben zich de afgelopen eeuwen voorgedaan en ook vandaag de dag bezwijken er dammen en dijken onder de druk van het water - met alle gevolgen vandien. In Nederland leidde de stormvloed van 1953 waarbij 1850 mensen omkwamen, tot de aanleg van de fameuze Deltawerken. Een recent voorbeeld van het bezwijken van waterkeringen betreft de orkaan Katrina in augustus 2005 die aanzienlijke overstromingen tot gevolg had met 1300 doden en tientallen miljarden aan economische en sociale schade.

De verwachting is dat de gevolgen van klimaatverandering de komende decennia zullen leiden tot meer bedreigende omstandigheden gedurende de levensduur van waterkeringen wat aanleiding is tot toegenomen aandacht voor veiligheid. Immers, als een waterkering het begeeft, dan kan dat grote waterstromen veroorzaken die een gevaar vormen voor mens en haard benedenstrooms. Eventueel verlies aan mensenlevens hangt af van waterdiepte en stroomsnelheid, waarschuwingstijd en de aanwezigheid van bewoners ten tijde van een eventuele ramp. Tijdige waarschuwing speelt een belangrijke rol bij het voorkomen van verlies aan mensenlevens. Het ontwikkelen en verbeteren van modellen voor risicobenadering en het afwenden van gevolgen van overstromingen, wordt steeds belangrijker bij het toetsen van de sterkte van reeds bestaande waterkeringen.

Het doel van dit onderzoek is om een raamwerk op te zetten en technieken te ontwikkelen voor het modelleren van het bezwijken van waterkeringen. Ook worden mogelijke nieuwe benaderingen op het gebied van risico analyse onderzocht. Naast numerieke en statistische benaderingen zijn 'constraint based' technieken ontwikkeld voor het modelleren van dijkdoorbraken en het afwenden van overstromingsgevaar. Er is een nieuw model voor dijkdoorbraak (BREADA) ontwikkeld voor het geleidelijk bezwijken van waterkeringen ten gevolge van het 'over'stromen. Dit is gevalideerd op basis van beschibare veldmetingen. Om de nauwkeurigheid van eerdere modellen na te gaan en om eventuele nieuwe informatie aan de beschikbare meetgegevens te onttrekken, zijn 'data mining' technieken gebruikt die al geruime tijd in de hydroinformatica met succes zijn beproefd op toepassingen in waterbouw en waterbeheer.

Op fysica gebaseerde modellen vragen om een goed begrip van onderliggende processen om een wiskundige beschrijving te kunnen opstellen; deze wordt vervolgens getoetst aan de hand van beschikbare meetgegevens. Data mining technieken gaan uit van dezelfde meetgegevens, maar proberen het fysisch proces hieraan te onttrekken zonder vooraf een bepaald model op te leggen. In dit onderzoek worden data mining technieken gebruikt om de maximale waarde van uitstroom bij dijkdoorbraak te voorspellen. Ondanks de beperkte hoeveelheid gedocumenteerde gegevens laat dit onderzoek zien dat het mogelijk is om de bestaande empirische modellen te verbeteren.

Een adequate analyse van een mogelijke catastrofale gebeurtenis als een grote damdoorbraak is essentieel om noodmaatregelen voor te bereiden die de benedenstroomse gevolgen van een ramp kunnen beperken. In dit proefschrift wordt een analyse uitgevoerd naar de een eventuele doorbraak van Bovilla Dam bij Tirana in Albanie, en worden mogelijke maatregelen onderzocht om een ramp af te wenden. Resultaten van BREADA en andere modellen worden gebruikt om de variatie in uitstroomcondities na te gaan. Met WL | Delft Hydraulics' Sobek 1D2D numerieke modelsysteem worden overstromingen gesimuleerd en gevoeligheidsanalyses uitgevoerd om onzekerheden vast te stellen. Voor mogelijk bedreigde gebieden worden noodmaatregelen onderzocht.

Het is duidelijk dat het falen van waterkeringen levensbedreigend kan zijn met grote economische gevolgschade. Daarom is het noodzakelijk om niet alleen een faalkans vast te stellen, maar ook om mogelijke gevolgen te kunnen beperken. Immers, mocht een faalkans worden overschreden, dan kunnen de gevolgen verstrekkend zijn, zoals bijvoorbeeld bij New Orleans. In dit proefschrift wordt daarom een alternatieve route bewandeld met de nadruk op het verkleinen van de gevolgen van overstromingen. Daartoe is een numeriek model gecombineerd met een 'constraint' aanpak gebaseerd op 'graph theory' wat het mogelijk maakt om meerdere scenario's in zeer korte tijd te evalueren. De resultaten van een toepassing in een gebied met meerdere polders met verschillende doelfuncties zoals bergingscapaciteit en economische waarde (ook wel systeemwerking genoemd) zien er veelbelovend uit.

De benadering die in dit proefschrift is gekozen kan gebruikt worden als aanvulling op de huidige aanpak die veelal is gebaseerd op het simuleren van een groot aantal mogelijkheden waarvan pas later blijkt of deze relevant zijn of niet. De hier voorgestelde 'lichte benadering' van gecombineerde 'numerical–constraint based' technieken is flexibel genoeg om zeer veel mogelijkheden in korte tijd te evalueren en om de meest waarschijnlijke scenario's te selecteren die vervolgens met behulp van een conventionele hydrodynamische aapak in meer detail kunnen worden onderzocht. Op deze manier wordt de rekentijd sterk gereduceerd en alleen gebruikt om de meest waarschijnlijke opties te onderzoeken. Een combinatie van beide methoden is mogelijk door bestaande hydrodynamische pakketten uit te breiden met opties voor een 'numerical–constraint based' benadering, of vice versa.

Acknowledgment

This dissertation would not have been possible without the continuous encouragement of my promotor Prof. Arthur Mynett. I am very grateful to him not only for directing my research, but also for the invaluable moral support I received throughout this project. His guidance and supervision enabled me to complete my work successfully.

I would like to acknowledge WL | Delft Hydraulics for their financial support and for providing the professional environment to carry out this research. Many thanks go to all my friends and colleagues at UNESCO-IHE, Delft University of Technology and WL | Delft Hydraulics: your friendship and professional collaboration meant a great deal to me. I am grateful to Dr. Hans Goossens and Dr. Henk van den Boogaard of the Strategic Research Department of WL | Delft Hydraulics for contributing their time and expertise to this project. Special thanks go to Prof. Roland Price and Prof. Dimitri Solomatine of UNESCO-IHE for the informal and fruitful discussions.

I would like to express my gratitude to the thesis committee members for their interest and valuable comments on my work and to Prof. Nigel Wright for helping improve the thesis with his useful observations and suggestions.

This list of acknowledgments would not be complete without all the people to whom I am indebted at a personal level. My friends and relatives have provided invaluable moral support during the four years of this research. I am very grateful to each and every one of you. I wish to express my special appreciation to my dear friend Merita Hatibi Serani who introduced me to the field of hydraulic engineering. Although she untimely passed away, the memories and creative energy she left behind are always with me.

I am as ever, especially grateful to my family. To my brothers for their love and encouragement. To my mother for her continuous support in my objective of enriching my knowledge, despite the pain of being away from her. To the happy memory of my father, who always believed in me but unfortunately did not live to see this thesis being completed, but who provides a persistent inspiration for my journey in life.

Migena Zagonjolli,

24 September 2007

Contents

Chapter 1

Introduction

If any one be too lazy to keep his dam in proper condition, and does not so keep it; if then the dam break and all the fields be flooded, then shall he in whose dam the break occurred be sold for money, and the money shall replace the corn which he has caused to be ruined.

The Code of Hammurabi (18th century BC)

1.1 Background

The history of water defence and water retention structures coexists with the history of their failures. Around the world thousands of dams have been constructed over many centuries. But also, hundreds of dams have failed and every year many dikes breach due to high flows in the rivers, sea storm surges, etc. often leading to catastrophic consequences. By far the world's worst dam disaster occurred in Henan province in China, in August 1975, when the Banqiao Dam and the Shimantan Dam failed catastrophically due to the overtopping caused by torrential rains. Approximately 85,000 people died from flooding and many more died during subsequent epidemics and starvation; millions of residents lost their homes (Qing, 1997). This catastrophic event is comparable to what Chernobyl and Bhopal represent for the nuclear and chemical industries (McCully, 1996). In the Netherlands, in February 1953, a high–tide storm caused the highest water levels observed up to date and breached the dikes in more than 450 places, causing the death of nearly 1,900 people as well as enormous economic damage (Gerritsen, 2005).

During recent years, the observation is made that the global impact of climate change will be devastating (see e.g. Intergovernmental Panel on Climate Change (IPCC) assessment reports at www.ipcc.ch; Millennium Ecosystem Assessment (2005); Archer (2006); Flannery (2006); Wentz et al. (2007)). According to the IPCC, "the observed warming trend is unlikely to be entirely natural in origin". We witness the

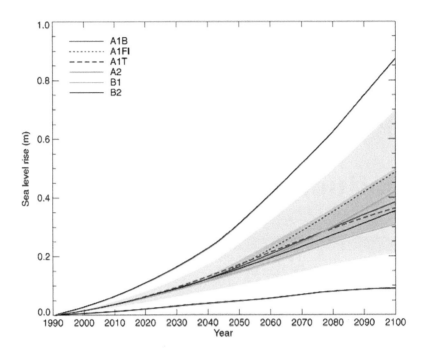

Figure 1.1: Global average sea level rise for the period 1990 to 2100 for different scenarios (IPCC, 2001).

rising of temperatures, which in itself does not always mean nicer sunny weather but more likely enhances the potential increase of the intensity of flash floods, heat waves, mudslides and droughts, leading to catastrophic social, environmental and economic damages. The declining ice extent in Arctic, the increase in melting rate of Greenland ice sheet, and rising global mean sea level (Figure 1.1) are the most evident consequences of global warming. The number of high category hurricanes has increased, while the first ever hurricane was recorded in 2004 in the South Atlantic (South Coast of Brazil).

The failure of the New Orleans' levee system during hurricane Katrina of 2005 contributed to prolonged flooding causing 1,300 casualties and billions of dollars of economic damage. In August 2002, floods caused by torrential rains in Europe claimed the lives of at least 109 people. The Elbe exceeded its 200 year flood return period in Dresden, flooding much of the city, and a flood return period of 500 years was estimated in Prague.

Nowadays there are more than 800,000 dams and thousands of kilometers of dikes, constructed around the world for different purposes: flood control (the most common purpose), irrigation, electricity generation, water supply, recreation, etc. Dams and dikes are often designed based on the statistical distribution of recorded river flows or sea storm surge measurements. However, climate change has introduced uncertainty related to increasing maximum flows within the life span of dams and dikes, leading to safety concerns. Many dams and dikes previously considered adequate (safe) now exhibit a potential to experience overflowing (overtopping) during high (flash) flood events. If a dam or dike fails, loss of life and economic damage are direct consequences of such an event, depending on the magnitude of water depth and velocity, warning time, and presence of population at the time of the event. Early warning is crucial for saving lives in flood prone areas.

Costa (1985) compared the loss of life during two dam failure events: the Teton Dam in Idaho (93m high) that failed at midday on 5 June 1976 and Laurel Run Dam in Pennsylvania (12.8m high) that failed on 20 July 1977 at 4:00 a.m. The latter event claimed the lives of 1 out of every 4 people exposed to floodwaters, while only one out of 3,000 people exposed to floodwaters suffered from casualties in the case of the Teton Dam failure. The database compiled by the Centre for Research on Epidemiology Disasters (www.em-dat.net) identifies floods as the second most endangering factor after drought in terms of loss of human lives around the world for the period 1900-2007.

When the failure of a tailing dam happens, the socio–environmental implications might affect the flooded region for a long period of time. The dam failure at the Los Frailes mine in Spain, in April 1998, released between 5 to 7 million tonnes of toxic mud carrying heavy metals and highly acid compounds in the Guadiamar River spreading in large areas of the floodplains. Large scale pollution of the Doñana World Heritage nature park, one of Europe's primary wildlife sites, was narrowly avoided thanks to an emergency dike constructed by the Spanish authorities. However, the environmental disaster was immense and its long–term consequences are feared to affect the region for many years to come (Olías et al., 2005).

The construction of dams and dikes (hereafter referred to as structures) leads people to believe that the floods are fully controlled, and therefore an increased urban and industrial development in the floodplains usually takes place. Hence, if the structure fails, the damage caused by flooding might be much greater than it would have been without the structure's presence. Having the historical failures of structures in mind, one might pose the question what can be done in order to reduce the risk posed from a dam or dike failure event.

1.2 Management of the flood risk caused by structural failure

The traditional approach of preventing the impact of flood through flood protection, is more recently being replaced by the flood management approach (Mynett and de Vriend, 2005; de Vriend, 2005; Samuels et al., 2005; Simonovic and Ahmad, 2005) as a result of the recognition that absolute flood prevention is unachievable and unsustainable, due to high costs and inherent uncertainties. The purpose of flood risk management is to protect the people facing risk up to a certain acceptable (affordable) level, *and to reduce the consequences of an extreme event exceeding the acceptable level*, so that a disaster can be avoided. Thus, risk management is defined to be the process of assessing and reducing the risk. Assessment of the risk involves recognizing the plausible failure modes for a structure or the plausible flooding events, quantifying probabilities and consequences (socio-economic and environmental) for all (or only plausible) failure modes or flooding events, and evaluating the risk by comparing the posed risk to the predefined risk criteria (acceptable or non acceptable risk level). Usually a threshold criterion is applied that establishes a level over which risk is considered unacceptable. International Commission on Large Dams (2005) defines the tolerable risk as follows:

> *A risk within a range that society can live with so as to secure certain net benefits. It is a range of risk that we do not regard as negligible or as something we might ignore, but rather as something we need to keep under review and reduce it still further if and as we can.*

The key approach in achieving tolerable risk is reducing risk as low as reasonably practicable. It is usually defined for each structure or system of structures (dike rings) rather than as a general criteria. It is based on a case specific evaluation of all possible risk measures, and different criteria might be applied for life safety and other consequences in different countries. A tolerable level of life risk is often evaluated in reference to risk of loss life due natural hazards or disease (Shortreed et al., 1995).

Different methods can be applied for reduction of the risk posed by a dam failure according to Bowles (2001):

1. Avoid the risk before or after the dam is built. If the safety of the dam is questionable, then decommissioning of the dam can be proposed as a solution to the problem.

2. Reduce failure probability occurrence through *structural* and *non–structural* measures.

3. Reduce (mitigate) consequences through transferring the risk, effective emergency evacuation planning or relocation of population at risk. The 'Peace

Dam' is constructed about 125 kilometers northeast of Seoul, South Korea for the purpose to mitigate the flood water in case of collapse of North Korea's Imnam or Mount Geumgang Dam or sudden release of water through the outlet works.

4. Retain (accept) the risk, but protection measures should be taken in the downstream area against the flood water as well as other measures should be applied to mitigate the flood water from the most populated or economically valuable areas.

The risk associated with flooding is generally expressed as the product between the probability of the event occurrence and the monetary value of its consequences. Expressing loss of life in monetary terms usually is not morally acceptable, thus, the population at risk is commonly taken into account as a decisive element in risk analysis. Risk expression might lead to the equivalence of an event with low probability and high consequences to an event that has high probability of occurrence but very low consequences. Generally, risk reduction measures try to reduce the probability of flooding, though minimizing the probability of a flood might come at the price of increasing its destructive power. The Indian Ocean tsunami of December 26, 2004 demonstrated that the consequences should be a triggering element of any risk analysis, instead of focusing on the probability of the event only. The impact of the Asian tsunami could have been lower in terms of human live losses if an effective warning system would have been implemented and operated.

The total failure probability of a structure involves the combination of individual probabilities estimated for different factors and loading conditions that contribute to its failure (Hartford and Baecher, 2004). Correctly defining the overall probability, while taking into account all factors that could lead to the failure of the structure is questionable, especially when extreme probabilities are deduced from small data samples. A new extreme flow record or new development at the areas potentially affected by the structure failure will make the calculated probability value outdated. Furthermore, in most cases the failure modes are not independent and therefore the failure probabilities are not simply additive.

Nowadays, as a result of economic development and population growth in flood prone areas, the potential flood damage increases as well. Furthermore, climate change caused by global warming might lead to more devastating flooding than ever. Higher river dikes are now seen more as a contributor to major flooding than a protection against it, in addition to having negative effects on nature, landscape and cultural heritage of the surrounding areas. In the Netherlands, the standard policy of raising dike crest levels in order to maintain the required level of flood protection is being abandoned in favour of the 'Room for the River' policy: widening river cross sections by relocating dikes further away from the river and/or by lowering the river floodplains, constructing floating houses, etc. (van Schijndel, 2005).

Unfortunately, we might fail to stop an event from happening due to our incapability, negligence or unpredictability of extreme natural events, but we can develop models and tools for fast response to any event in order to reduce its consequences. The aim is to find measures that reduce the probability of flooding and minimize the potential consequences.

1.3 Scope of the thesis

Developing and improving flood propagation, risk assessments and flood mitigation models for already constructed dams and dikes is becoming a necessity for a variety of reasons such as decreasing human casualties and economic damage. In this thesis, instead of focusing on methodologies to estimate and lower the failure probability of hydraulic structures, we propose approaches that cope with hazards caused by structural failure events by decreasing their consequences. We consider events, though not likely to happen in any given year, if occurring are extremely catastrophic and have enormous socio–economic impact.

We address the problem of dam and dike breach analysis as well as simulation and mitigation of the flood water caused by failure of these structures. Formation of the breach in a structure is a complex process that depends on various hydraulic, hydrologic, geotechnical factors. In this thesis, we develop a framework and techniques for modelling dam and dike failure events as well as propose several novel approaches for dam breach modelling. Furthermore, we introduce and apply several numerical, statistical and constraint based methods in particular related to dam and dike breach modelling and flood water mitigation.

In this thesis, a 'lightweight' numerical–constraint based technique is proposed. This technique offers advantages of simultaneous evaluation of different flood mitigation scenarios. Through constraints we optimize the strategy for choosing the most feasible flood propagation scenario that minimizes economic consequences.

The objectives of this thesis can be summarized as follows:

1. Review and comparison of different existing methods (mainly physically and statistically based) for dam and dike failure modelling.

2. Development of a dam breach model for breach formation in earthfill dams and its validation against a real dam failure event.

3. Development of a methodology for predicting dam breach characteristics and peak outflow during the failure event by using statistical and data mining techniques.

4. Development of an approach for reduction of flood consequences caused by a dam failure event, identification of areas prone to flooding, identification of

uncertainties (input, model, and completeness uncertainty), and proposal of risk reduction measures.

5. Development of a numerical constraint based model for flood mitigation in low–lying areas subject to flooding.

1.4   Outline of the thesis

The thesis is composed of eight chapters. Chapter 2 presents a literature review on earthfill dams and dikes, focusing on their failure analysis and the processes involved during the breaching of a structure.

Chapter 3 describes in detail current state–of–the–art approaches used for dealing with floods in particular in the Netherlands and Japan. Furthermore, it provides in depth overview of the applied methods, namely their characteristics, advantages, and disadvantages.

Chapter 4 compares the theoretical and practical aspects of available approaches and/or models used for breach modelling of water defence and water retention structures.

Chapter 5 introduces a new approach for the estimation of dam breach characteristics. It starts with an overview of the theory behind the data mining techniques used in this research and demonstrates their application to the dataset of dam failure events.

Chapter 6 presents the mathematical model of a dam breach tool developed during this research as well as its application to the modelling of a hypothetical failure of an earthfill dam. It provides an evaluation of breach characteristics using different approaches, a sensitivity analysis of important breach parameters and an evaluation of uncertainties.

Chapter 7 describes the development of a numerical–constraint based model which is capable to simulate different flood mitigation scenarios taking into account the social and economic value of areas that could be prone to inundation.

Chapter 8 presents the conclusions of the research and future recommendations.

Chapter 2

Water Retention and Flood Defence Structures

Engineering is the professional art of applying science to the optimum conversion of natural resources to the benefit of man.

Ralph J. Smith

2.1   Design criteria and failure modes for dams

A dam is a barrier made of earth, rock, or concrete or a combination thereof that is constructed across a river for impounding or diverting the flow of water. The history of dam construction dates back to 2900 B.C. with the oldest dam in the world believed to be constructed in Wadi el–Garawi, 30 km south of Cairo, Egypt (Singh, 1996). The dam was built for irrigation purposes and had a crest length of 106m and a maximum height of 11.3m. It collapsed the first winter it was in use, but its remains are still present today. The Alicante Dam in Spain was 46m high when completed in 1594 and remained the world's highest dam for 300 years. Currently, hundreds of high dams are in operation worldwide. The Nurek Dam in Tajikistan (300m high) is the highest and the Three Gorges Dam in China with a reservoir storage of 39.3 billion m^3 is the largest dam in the world.

The construction of dams is often seen as a solution for providing water supply, flood control or for 'green', renewable electricity. The dams must be designed, built, and operated so that they make a positive contribution to socio–economic development, while having minimal impact on the environment. There are, however, different perspectives on this issue and discussions are taking place about the controversial impacts of dam construction. Resettlement of people that have to withdraw from their social and cultural identity is often seen as a major impact of dam construction, especially for large dams. Irreversible degradation of natural habitat, degradation of water quality, sedimentation of reservoirs and the downstream effects are other negative impacts of dam construction. The reader is referred to the report by the World Commission on Dams (2000) for a detailed description of the complex impacts of dams. To alleviate the impacts, measures are usually taken for adequate mitigation of the natural habitat e.g. by creating nearby protected areas for wildlife, or new fisheries within reservoirs, etc. The dam proponents propose more hydropower dam construction to play "a major role in reducing greenhouse gas emissions in terms of avoided generation by fossil fuels" (Lafitte, 2001). On the opposite side, environmentalists insist that any large dam, including hydropower dams, emits greenhouse gases (GHGs) due to the rotting of the flooded organic matter. However, the science of quantifying GHGs reservoir emissions is uncertain and controversial conclusions are drawn related to hydropower emissions in comparison to those from fossil fuels, viz. the difference between the pre–dam emissions from the undamed catchment and the post–dam emissions. In this thesis we focus on a dam as a structure, while its construction impacts - though very important - are out of the scope of this thesis. The interested reader is referred to e.g. Galy-Lacaux et al. (1999); Rosa and dos Santos (2000); Soumis et al. (2005); Tremblay et al. (2004).

Constructed dams can be categorized in two large groups: gravity and arch dams (Figure 2.1). Gravity dams rely on their weight to resist the forces imposed upon them. Arch dams, with the arch pointing back into the water, use abutment reaction forces to resist the water pressure force. They can be made of concrete or masonry. Gravity dams can consist of concrete or earth, rock, a mixture of these materials, or masonry. Dams are designed to have a low probability of failure during their construction and operation life span. Dam design criteria require the dams to withstand different loads, namely construction and reservoir water load, with or without seismic load. The greater the chance of loss of life or damage to valuable property in case of the failure event, the safer the design should be. Despite this, dams do frequently fail.

The failure of a structure can be partial or complete. The failure of the structure to fulfill its purpose is another type of failure. In this thesis, the term 'dam failure' indicates the partial or complete collapse of the dam or its foundation, leading to uncontrolled release of water in the downstream areas. Landslides in the reservoir might cause the release of water in the downstream areas, despite no failure of the dam structure occurs. A wave estimated to be 100m high overtopped the Vaiont arch dam in Italy in 1963 when a massive rockslide of 240 million m^3 fell into the reservoir at a velocity of approximately 30m/s (Pugh and Harris, 1982). Only minor damage to the dam crest was observed though the wave that reached a height of 70m at 1.6km downstream the dam caused the loss of 2,600 human lives. Failure of a dam can be sudden or gradual. A sudden failure is associated with concrete

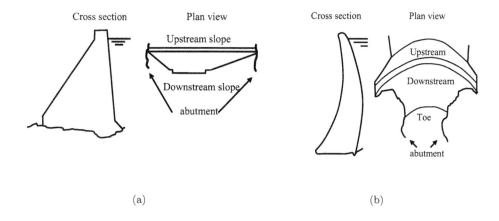

Figure 2.1: Schematic picture of (a) gravity and (b) arch dams.

dams, gravity or arch dams*. If breaching is initiated, the further development is faster than for earthfill dams under the same conditions. This observation allows modelling of the concrete dam failure events simply as a sudden (gate opening) process. Rockfill and earthfill dams, termed embankment dams, constitute the largest percentage of constructed dams around the world and not surprisingly the largest number of dam failure accidents occur with these dams, particularly with earthfill dams. Their failure, depending on the triggering factors, is mostly a gradual process rather than a sudden one. In this thesis we investigate the failure of embankment dams, in particular of earthfill dams.

Failure of an earthfill dam can be triggered by different factors, e.g. overtopping, foundation defect, seepage and piping. Overtopping is one of the most common failure modes for earthfill dams. It can be triggered by inflows higher than the design inflow, malfunctioning or a mistake in the operation of the spillway or outlet structure, inadequate carrying capacity of spillways, settlement of the dam or as a result of landslides into the reservoir. According to National Performance of Dams Program in USA (NPDP, 2007), 245 of 256 dam failure events recorded in the USA during the year 1994 happened due to high inflow discharges. Any embankment dam will fail if the spillway capacity is too small and flood waters rise high enough to flow over the top of the dam for a considerable amount of time. In August 1979, a flood two to three times larger than the design flood triggered the failure of the Machhu II dam in India, causing more than 2,000 casualties (Hagen, 1982). Once

---

*The Malpasset Dam in southern France, an arch dam of 66.5m in height and maximum designed reservoir capacity of 55 million m^3, is described to have failed *explosively* on 2 December 1959 (Hervouet, 2000).

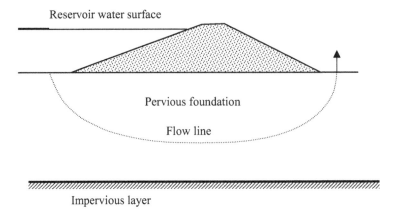

Figure 2.2: Seepage on the foundation of a dam.

an initial breach channel is created, and the high reservoir water levels persist, the breaching will continue to develop and any effort made to stop it will be unsuccessful. Overtopping may not result in structural failure, but still presents a major flood hazard as happened during the overtopping of the Vaiont Dam in Italy (Pugh and Harris, 1982). Similarly, rapid release of reservoir water in order to lower the water level within safe limits, can be a big concern in downstream areas.

Water penetrating through the dam's interior body or its foundation might progressively erode soil from the embankment or its foundation leading to the failure of the dam. Here, we define piping failure as a failure mode caused by water penetrating through the dam's body, carrying with it small particles of dam material, continuously widening the gap. If the initial piping can be detected before it reaches the critical condition, remedy might be possible. Penetration of water in the dam body can cause slope failure. To prevent this type of failure, appropriate instrumentation is needed to estimate the rate of infiltration within an embankment.

Seepage failure (Figure 2.2) or foundation failure occurs due to the saturation of the foundation material leading to either washout of the material or a weakening of the rock towards a sliding failure. The flow of water through a pervious foundation produces seepage forces as a result of the friction between the percolating water and the walls of the pores of the soil through which it flows. Figure 2.2 shows how water flows through the pervious foundation of a dam.

Earthquakes or sabotage are yet two other causes of dam failure. Earthquakes that have stimulated immediate failure of a dam appear to be very rare. The upstream slope of the Lower San Fernando Dam in California (USA) failed due to liquefaction during the earthquake in 1971. The dam was constructed by fill soil mixed with a large amount of water, transported to the dam site by pipeline, deposited on the embankment in stages, allowing the excess water to drain away. The fill that remained was loose, and was subject to liquefaction as a result of the earthquake. Fortunately, the reservoir level was low at the time of the earthquake and no flooding occurred. Failure due to an earthquake might result in a higher threat to the population downstream rather than the overtopping failure. In the first case a sudden breach of the dam would cause a flood wave moving downstream while the population might have no clue about the structural failure of the dam and the approaching flood water. On the other hand, prior to an overtopping failure, rising flood waters often give reason for concern to the residents in the floodplain area and lead to issuing flood warning.

The historical database of dam failure events shows that the number of failures caused by sabotage is small. One example is the British bombardment of the German dams on the Ruhr River during World War II. The Dnieprostroy Dam (43m high) on the Dnieper River was also destroyed for the purpose of preventing the movement of German troops.

2.2   Design criteria and failure modes for dikes

A dike is a barrier built along the shore of a sea or lake or along a river with the objective of holding back water and preventing flooding (Figure 2.3). Dikes are often constructed in the floodplain for the purpose of protecting from flooding. They are usually built from sand, clay or a combination of them, or from peat (like in many places in the Netherlands, where local soils were used for construction).

There are many aspects that are important in making decisions related to dike construction. Economic, environmental and other social interests have to be considered, involving different parties in the process of decision making. Different methods exist for the design of flood defence systems or water impounding structures: *probability* and *risk–based* design methods.

In a probabilistic approach, dikes are designed based on a water level with a particular frequency of being exceeded. The design flood levels as well a safety level or a margin ensure the dikes' integrity. Depending on the probabilistic method used in the analysis and on the available data records, the design levels might be different. Thus, including uncertainty in the estimation of the design levels is necessary, and the margin can deal to some degree with the small uncertainties. During the years the design levels might change as a results of changing flow conditions or as a result of development in the area protected by dike. Therefore different measures are implemented to ensure the structural integrity that take into account new developing conditions.

Dikes might fail due to different reasons and triggering factors. The common failure modes that are similar to the dam failure modes are piping, seepage, overtopping

Figure 2.3: Dike along a (a) coastline, (b) river, (c) channel, and (d) on the flood-plain.

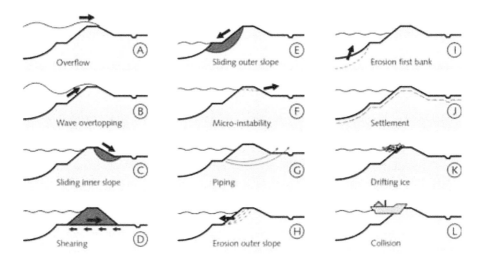

Figure 2.4: Possible failure modes for dikes (TAW, 1998).

Figure 2.5: Failure of a canal dike in Wilnis, the Netherlands (2003).

(overflowing), uplift, and slope failure. For a typical dike, various potential failure mechanisms are distinguished as shown in Figure 2.4. Water is not the only one triggering factor of dike failure. In August 2003, after a very dry and warm summer, one section of a peat dike (Figure 2.5) constructed along a canal in the Netherlands suffered a horizontal sliding, causing flooding of about 600 houses and evacuation of 2,000 people. In specific cases, the integrity of a dike might be affected by the activity of burrowing animals (field mouse, gopher, mole and fox). In case of their presence, actions are usually taken to control their activity through wire screening or traps placing along the structure.

The dike failure probability is calculated on the basis of a so–called *reliability function* Z. If the resistance R is determined and also the load S is known, the determination whether the structure will fail is simply $Z = R - S$. Here load and strength are both stochastic variables. If we consider the stochastic nature of the strength and load, the probability that the construction will fail is the probability of $P\{Z < 0\}$. If $Z = 0$, the *limit state* is reached, which constitutes the failure boundary.

Calculating the failure probability is as complex as modelling the dike failure processes. The factors affecting accurate estimation and modelling are:

- Variation of the properties of structural material (strength, deformation, permeability, time–dependent consolidation, compaction of the material, different quality of construction works, length of the dike).

- Lack of a reliable mathematical description of the failure processes needed for accurate determination of the limit states. As for overtopping and overflowing, these processes are understood with some degree of accuracy, but much less is known for other failure modes, such as piping.

- Uncertainty in estimating the overall failure probability of the dike. The total failure probability of a structure involves the combination of individual probabilities estimated for different factors and loading conditions that contribute to its failure. Correctly defining the total probability, while taking into account all the factors that could lead to the failure of the structure seems questionable. In most cases the failure modes are not independent and therefore the failure probabilities are not simply additive.

For the probabilistic safety analysis to be meaningful the accurate computational models and sufficient (statistical) data are needed. Different techniques are available for determining the failure probability given a reliability function and statistical characteristics of the basic variables. The uncertainties associated with them "are often even greater than the uncertainties due to the intrinsically stochastic character of load and strength" (CUR/TAW, 1990). As a result, the calculated value of failure probability is often used as a relative indicator for implementing priorities in maintenance of the dike ring system. For detailed analysis the reader is referred to CUR/TAW (1990) and Thoft–Christensen and Baker (1982).

Due to the shortcomings of the probabilistic approach (CUR/TAW, 1990), the risk analysis approach is drawing more attention. The risk based design approach considers the probability as well as the consequences of inundation in case of a failure of a flood defence system. Breaching of the dikes and consequent flooding have claimed many lives and caused enormous economical damages worldwide. Based on the risk estimation method the magnitude of the damage or loss is considered during the design of the dike. For detailed description see CUR/TAW (1990) and Vrijling (2001).

In this thesis we only consider the overflowing and overtopping failure modes that are the most common failure modes for embankment dams and dikes. In the following section we give a general description of the breach development characteristics and emphasize the differences between modelling of the breaching processes in earthfill dams and dikes.

2.3   Breach modelling

Breaching of a structure is a time–dependent and non–linear phenomenon. Water–soil interaction together with often non–homogeneous and specific material properties for each structure lead to the difficulty of accurately modelling the processes involved in breach development. Hydrodynamics, sediment transport mechanics, and geotechnical aspects are all present in the breach formation and their accurate modelling is very important for the accurate prediction of breach outflow. The development of effective emergency action plans and the design of early warning systems heavily rely on these prediction results.

There are similarities and differences in the processes involved during the breaching of embankment dams and dikes, as summarized below:

- Dikes in comparison to embankment dams are typically longer than high. While the breaching of a dam might develop to the limit of its geometry, for dikes only one section is usually breached. Also, the settlement along the dike length might vary in different sections, resulting in different crest elevations along the dike length.

- The breaching of embankment dams depends mostly on reservoir volume rather than river inflow. During the breaching of river dikes and sea dikes, the river flow and sea surge determine the breach development respectively. A relatively finite volume of water is involved in dam breaching and river dike breaching, opposite to sea dike breaching where an infinite and periodic volume of water is caused by tides.

- The flow in the river is parallel to the river dike axis, while for embankment dams and sea dikes the flow direction is perpendicular to the structure axis.

- Hydraulic load behind a dam is usually larger than the one for a dike.

Numereous methods have been developed for the purpose of modelling breach development as further discussed in Chapter 4. Here we elaborate on specific breach characteristics and the processes involved during breaching.

2.3.1   Breach shape

The mathematical description of the interaction between dam material and water flow is not yet fully accurate (Morris, 2005). The opening formed in the structure during the failure process - from here on defined as breach shape - depends on that interaction. For accurate modelling, the soil mechanic parameters should be known. Nevertheless, they can be determined only with limited degree of accuracy. Assumptions are made concerning the breach shape in order to avoid the non–linearity in the equations. Models usually predefine the shape of the breach. Constant breach shape and uniform erosion of the breach section throughout the whole breaching development time, is usually assumed. The breach cross section is often considered to be triangular, rectangular, trapezoidal or parabolic (Figure 2.6).

Johnson and Illes (1976), after analysing the data from approximately 100 case studies concluded that the breach develops initially in 'V' shape, three to four times

Figure 2.6: Erosive patterns of various breach shapes.

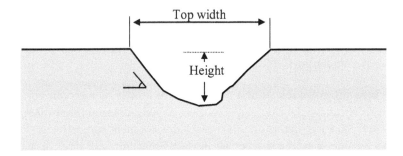

Figure 2.7: Parameters of the breach opening shape.

wider than deep, later developing in the lateral direction, once the apex reaches the hardest material of the dam core or its foundation. The lateral erosion continues until either the dam is completely washed out, or the reservoir is emptied. MacDonald and Langridge-Monopolis (1984) observed from the collected historical dam failure events that in most of the cases the ultimate breach shape is of trapezoidal shape. They concluded that for embankment dams, the breach shape can be assumed to be triangular up to the time that the base of the embankment is reached. Once the apex of the triangle reaches the foundation level, the breach develops forming a trapezoidal section extending due to lateral erosion. The conclusions drawn from several field and laboratory tests performed within the IMPACT project is that breach sides preserve the vertical angle during the breach development (Morris, 2005). However, a factor influencing the conclusion can be that the rectangular initial breach shape is predefined in all experiments. Data related to the progressing breach shape development (in time) during real dam failure events are still missing.

The parameters that specify the shape of a breach channel are: the breach depth h_b or the vertical extent of the breach measured from the dam crest down to the breach bottom, width at the top B_t and bottom B_{bot} of the breach channel, and the breach side slope factor z (see Figure 2.7).

2.3.2   Breach development

Breach development in time

Breach initiation time is defined as the time of duration starting with the first observable flow over or through the structure that might initiate warning, evacuation, or awareness, and ending with the start of the breach formation phase. During the breach initiation phase, the outflow is relatively small, and if it can be stopped the structure might not fail. Typical breach initiation times may range from minutes to days. Especially piping failure might be preceded by a prolonged initiation phase. The breach formation phase is considered to begin at the point where the structure failure is imminent and ends when the breach has reached its maximum size. For small reservoirs, the peak outflow from a dam break may occur before the breach fully develops due to significant drop in reservoir levels during the formation of the breach, whereas in larger reservoirs the peak outflow may occur when the breach has reached its maximum size. During the breach formation phase, outflow and erosion are rapidly increasing; while for a dike it might be possible to stop the breaching, it is unlikely that the outflow and failure can be stopped in case of an embankment dam. Several small springs were noticed near the right abutment of the Teton Dam, one day before its failure. All efforts made to close the sinkholes while the leak was rapidly growing, failed. In contrast, dike breaching can be stopped by human intervention; the famous Hans Brinker story[†] is a typical example.

The rate of breach formation depends on soil material properties (cohesive, non–cohesive, compaction, etc.) and embankment condition. Breach formation in embankment dams is highly dependent on the reservoir capacity and continues till either the reservoir is depleted or the dam can withstand further. According to the historical data, the breach formation phase for embankment dams ranges from 0.1 to 4 hours. The breach formation in the dike structure depends on the river flood or sea storm conditions as well as on the dike material. Cohesive dikes are likely to breach slower than non–cohesive dikes.

Breach development in space

All breach models assume an initial channel to have been created on the structure body (Figure 2.8) either parallel to the structure crest (DEICH_A (Broich, 1998)), with rotation around the downstream toe, or parallel to the downstream face (BREACH (Fread, 1988), BRES (Visser, 1998)). The initial channel forms the starting condition for the breaching process. If no initial channel exists then the subsequent stages of the breaching process will not occur. The initial characteristics of the breach channel define further breaching development.

[†]The legend tells the story of the brave Dutch boy thought to be named Hans Brinker, who supposedly prevented the flooding of Harlem city by pressing his finger in the dike opening (Dodge, 1997).

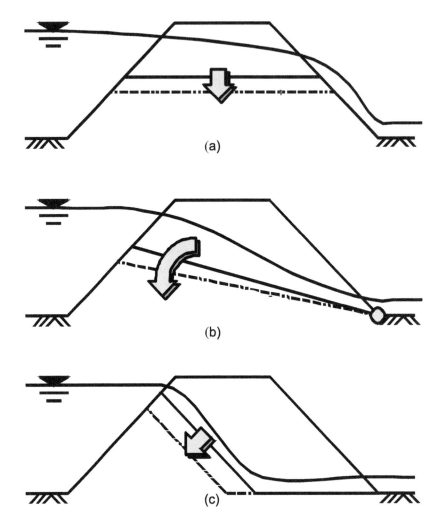

Figure 2.8: Modelling modes of breach growth: (a) parallel to the structure crest, (b) with rotation around downstream dam toe, and (c) parallel to downstream face (Broich, 1998).

It is rather difficult to predict where the initial breach channel will be formed as this depends on many factors e.g. structure coverage, flow characteristics, bad compaction at any point in the structure. Initial erosion may begin anywhere in the structure. For embankment dams, the developed models commonly assume a breach located at the center of the dam. However, some historical failure events showed that breaching might occur near an abutment as well. Examples include the failures of Teton Dam (Figure 2.9), Baldwin Hills Dam, etc. The breach development and outflow from a centrally located breach will most likely be different from a 'side'

Figure 2.9: Teton Dam failure (Rogers, 2007).

breach in terms of time to peak discharge, peak value, and hydrograph shape. When lateral growth is restricted in one direction, erosion rates in the other direction do not compensate (Morris, 2005). Therefore, to predict the initial breach location there is a need to undertake local surveys to identify weaknesses in the structure or sub surface geology by visual means, sensors or remote sensing techniques. Within the scope of this thesis, no attention is given to the processes that might define the development of the initial channel location.

The structure might be homogeneous or heterogeneous with an impervious core at the center of the structure (embankment dams). Different types of material may lead to different channel slopes. When breaching is initiated at the downstream face of the dam, a steep slope can be observed at the first stage of dam breaching due to overtopping, but as the breaching continuous, the slope might remain about the same or even decrease.

2.3.3   Breach formation mechanisms

Two breach formation mechanisms are identified: erosion and headcut erosion. The latter is the process of removal of structural material by the combined effect of the erosive force of water flow and by mass wasting (see Figure 2.10). Laboratory experiments and observations of real earthfill structure failures show that erosion is predominant for non–cohesive structures without a cohesive core. Headcut erosion is observed to be predominant during the breaching of structures with cohesive filling material, or with non–cohesive filling material but with a cohesive core (see

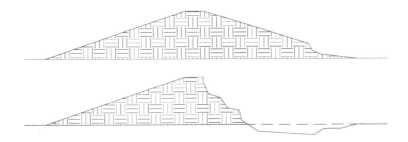

Figure 2.10: Headcut erosion process in a cohesive soil embankment.

e.g. Ralston (1987); Dodge (1988); Powledge et al. (1989); Hanson et al. (1999); Morris (2005). By validating the modeling results versus field and laboratory experiments carried out during the IMPACT project, the breach models that predict breach growth considering the headcut erosion processes rather than only erosion, were argued to perform better (Morris, 2005) than the models that consider only erosion.

Modelling of the headcut erosion is not trivial, and while many experiments are carried out to gain insight into this process, the mathematical modelling of this process is just at the initial stages (see e.g. Temple and Hanson (1994); Temple and Moore (1997); Wu et al. (1999); Robinson and Hanson (1994); Hanson et al. (2001); Alonso et al. (2002)). Most breach models either do not consider headcut erosion, or consider this process using very simplified assumptions, usually modelling it as an energy dissipation process.

Erosion is modelled using the sediment transport equations that are conventionally derived for steady subcritical flow conditions, specific types and certain diameter ranges of sediment (Yalin, 1972; van Rijn, 1993; Bogárdi, 1974). During structure breaching, the flow might develop into unsteady, supercritical flow and if these conditions apply, the use of unsteady non–uniform sediment transport equations is more appropriate. However, due to their absence the Meyer-Peter and Müller (1948); Exner (1925), Einstein–Brown (Brown, 1950), and the modified Meyer–Peter and Müller formula adapted by Smart (1984) are commonly used.

The rate of erodibility is given by:

$$E_r = k(\tau - \tau_c)^a$$

where, E_r presents the erosion rate, k and a are two correlation coefficients, τ presents the flowing water tractive stress and τ_c is the critical tractive stress for the erodible material.

The rate of erosion is commonly assumed to be uniform throughout the channel

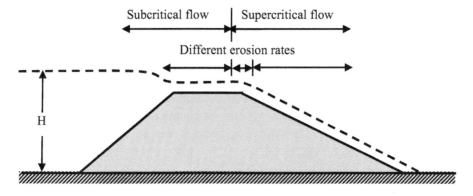

Figure 2.11: Description of the flow over an embankment as from Powledge et al. (1989).

section that is the submerged part of the breach channel sides is supposed to erode at the same rate as the unsubmerged part and the breach channel bottom. Most of the mathematical models deal with the homogeneous dams or dikes. The failure modelling of the heterogeneous structures is commonly done through averaging the characteristics of soil properties.

2.3.4 Hydraulics of flow over the dam

The breach outflow hydrograph is of crucial importance for the assessment of the flooding characteristics in the downstream areas. The available breach models simulate flow through the breach channel using either the orifice equation (at the initial phase of piping failure) and the weir equation, or the one dimensional de Saint–Venant equations.

The weir equation that estimates the unit discharge for the free flow (low tailwater) condition is generally expressed as:

$$q = CH^{1.5} \qquad (2.1)$$

where q is the discharge per unit width, C is the discharge coefficient that depends on the weir (breach) shape, and H is the total (energy) head above the crest. The coefficient C ranges between 1.60 and 2.15 in SI metric units. If the crest is submerged by tailwater, then Eq. 2.1 becomes

$$q = C_s H^{1.5} \qquad (2.2)$$

where, C_s is a coefficient that considers the submergence effect. Powledge et al. (1989) describe the three flow zones (Figure 2.11) observed during overflowing of the dam with no tailwater effects. The first zone is the movement of the flow from the calm reservoir (static energy head) to subcritical velocity state (static and dynamic head) over the upstream portion of the dam crest. In this zone the hydraulic forces and flow velocities are low. The small energy slope of the subcritical flow range imposes small tractive stresses too. In the second zone, the flow travels through critical velocity on the crest to supercritical flow across the remainder of the dam crest, to the downstream slope. In this zone the tractive stresses might become significant. In the third zone a rapidly accelerating turbulent supercritical flow is observed on the steep downstream slope. Here the energy levels increase significantly as the flow proceeds along the downstream slope of the dam. Due to the steep energy slope, the vertical velocities might increase significantly and the tractive stress will be large. The downstream slope of the dam is a steep slope in hydraulic terms and a correction must be made in the calculation of gravitational forces.

The weir formula is often used to calculate the flow along the crest. The steady non–uniform flow equations have also been used to compute the water depths, velocities, and energy slope on the downstream slope despite the short reach of the breach channel and its steep slope because of their relatively simpler computations compared to full de Saint–Venant equations.

Chapter 3

Current Approaches for Dealing with Flooding

Happy Holland, had we not dug and even diked,
We were now living above the rivers,
That must cut the land, but now run over it in man–made
channels,
Whose bottom is ever higher raised by falling silt,
And so rise ever more above the land and in force and violence
their dikes can overwhelm.

Willem Bilderdijk (XIX Century)

3.1 Introduction

Nishat (2006) defines flood as "the process of inundating normally dry areas and causing damages". Inundation does not always pose a risk and not necessarily has only negative impacts. It can have positive effects on soil fertility, ecosystems, etc. (Nishat, 2006). Floods*, on the other hand, can be very severe and have enormous consequences in terms of economic, ecological and social values. Floods' destructive forces are a threat that has been faced by humans for generations already. Noah's legend is one of the thousands of legends related to floods and their devastating consequences.

Therefore, novel strategies and methods are continuously developed worldwide for dealing with floods (Knight et al., 2006). Several common types of measures are distinguished, such as:

1. Measures for preventing flooding: dam or dike construction, maintenance and improvement, river dredging, etc.

---

*Reader is referred to FLOODsite (2005) for definition of flood and risk related terms.

2. Measures for reducing flood impact: retention and detention basins, floodways, flood forecasting, spatial planning, awareness raising (games, role–plays, brochures, etc.).

3. Measures for dealing with an approaching flood and during a flooding period: Decision Support Systems (DSS), warning and emergency plans, evacuation and local emergency protection, etc.

4. Measures taken after the flooding occurred: aftercare, compensation, insurance and restoring of the flooded area. The loss or damage of property, and in some cases the inability to return home for a period of time causes great stress and disruption to people. The metro of Prague, was unable to run for several months after the 2002 torrential flood that hit large parts of Europe. Measures should be taken to deal not only with direct but also with indirect (long term) consequences.

These measures taken against flooding are categorized in two groups of strategies:

1. Flood controlling strategies referred to as *resistance* strategies, and

2. Flood damage reduction strategies referred as *resilience* strategies (Vis et al., 2003).

The first strategy aims at fully preventing floods. However, as flood prevention is not always possible and feasible, nowadays more emphases is given to strategies that improve the coping capacity, resilience, and adaptability, and offer more flexibility for future interventions. Disaster mitigation is suggested as a priority at numerous international events such as the Fourth World Water Forum in Mexico[†] in 2004, the Conference of International Center for Water Hazard and Risk Management[‡] in 2006, etc. In the following sections we describe flood dealing strategies in two countries that are similarly vulnerable to water related disasters.

3.2 The Netherlands' long history of battle against floods

"...Dutch created the Netherlands", is the Dutch expression that emphasizes the continuous battle against water to expand or protect the Netherlands territory through which three major European rivers flow into the sea: the Rhine with its branches the Waal and the Neder Rijn flowing from Germany, and the Maas (a branch of the Meuse) and the Schelde, flowing from Belgium. Being one of the most densely populated deltas in the world with 25% of its land area below sea level and 65% prone

---

[†] Reader is referred to www.worldwaterforum4.org.mx
[‡] Reader is referred to www.icharm.pwri.go.jp/html/meetings/iwfrm2006

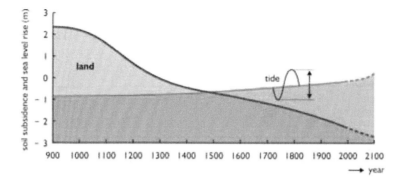

Figure 3.1: The land and sea level throughout the years in the Netherlands (Klijn et al., 2004). The dashed line presents the forecasted estimates.

to flooding, the Netherlands has no other choice of existence but to fight against sea and river high water levels. The land has been continuously subsiding (Figure 3.1) as a result of drainage during the Middle Ages' reclamation and the action of both shrinkage and oxidation of peat layers. Since the eleventh century, when the first dikes are known to have been built (van de Ven, 2004), nowadays, 53 rings of dikes protect the delta area from flooding.

In 1927, after the severe flood of 1916, the implementation of the Zuiderzee Works started (see Figure 3.2). A system of artificial dams, land reclamation and water drainage works was built. The Afsluitdijk or the 'closure dike' (in English) has been the longest man–made dam[§] for 74 years. Some 32km long, 90m wide and 19m high (some 7.25m above sea level), the dam is essential for protecting the land from the North Sea. After the completion of the works, the coastline length was decreased by almost 300km and a freshwater lake was created that provides drinking water as well as reclaiming land for agriculture and urban development.

During the February 1953 storm (Gerritsen, 2005), the greatest storm surge on record for the North Sea, the Afsluitdijk prevented flooding on the Zuiderzee coast. However, on the southwest of the country, the storm lead to the worst ever disaster in the Netherlands when 800km of dikes protecting the area were breached causing 1,835 casualties and flooding 2,000 km^2 of land. Afterwards, the construction of the Delta Works (Figure 3.3), the most spectacular water related mega structure in the world, took place.

Numerous dams and dikes, sluices, locks, and storm surge barriers were built in the Netherlands to protect the land from sea and river flows. However, in 1993 and 1995,

---

[§]The Saemangeum Seawall, located on the southwest coast of the South Korea, is the world's longest man–made dike from the time of its inauguration in 2006, being 0.5km longer than the Afsluitdijk dike.

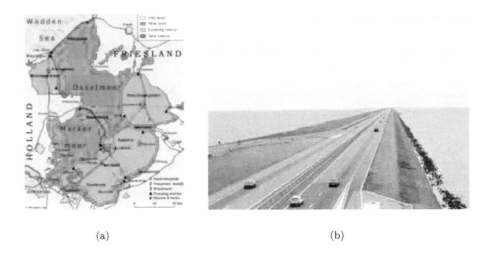

(a)   (b)

Figure 3.2: Afsluitdijk - the 'closure dike' - in the Netherlands.

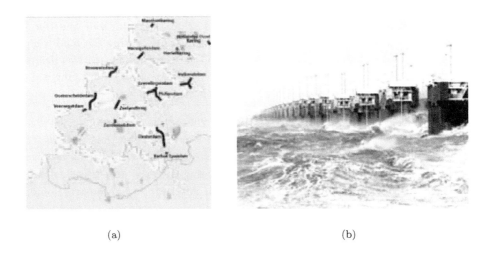

(a)   (b)

Figure 3.3: Eastern Schelde storm surge barrier (Rijkswaterstaat).

the Netherlands was again under threat of significant flooding, when the Rhine and the Maas Rivers reached very high levels and the stability of the dikes seemed no longer guaranteed. Some 200,000 people were evacuated during the January 1995 storm.

Nowadays, there are over 16,000km of dikes and 300 structures that protect the Netherlands from flooding. About 2,400km of constructed dikes are designated as primary dikes - dikes in direct contact with water bodies (sea, lakes and rivers) - and 14,000km are secondary dikes - dikes sorrounding the canals. The design of sea and river dikes is traditionally based on a design water level with a particular frequency of exceedance. Following the disaster of 1953, the Delta Committee (1960) defined the design water levels for the sea dikes, which are built to withhold the 10,000 year return period storm surge levels. The Commission on River Dikes (1977) recommended that river dikes should be designed and improved to retain water levels associated with a governing rate of discharge of the Rhine River at Lobith (entry location of the Rhine River in the Netherlands), which is exceeded with a frequency of about 1,250 year (CUR/TAW, 1990). The design levels keep increasing, as the current 1,250 year return period discharge at Lobith corresponds to 16,500m^3/s while further increase in the design discharge is forecasted to be from 16,800 (minimum scenario) to 18,000m^3/s (maximum scenario) by the year 2100 (Kwadijk and Rotmans, 1995). Still, those values are uncertain as the 1,250 year flood event is forecasted based on only 100 years of records.

Innovation in flood protection, prevention, and management is continuously developing as the perception that the Netherlands can be subject to flooding anytime has not perished. Climate change has increased the uncertainty about river flows while sea level rise might impede the rivers' discharge. Moreover, sedimentation that occurs on the floodplains along the rivers and the subsidence of the land behind the dikes, lead to larger differences between floodplain and hinterland level (Asselman and Middelkoop, 1995). In other words: when the river capacity is decreased, the flood hazard is increased. Construction of dikes leads people to believe that they are safe, and as such the urbanization and economic development of the areas continue to grow, increasing in turn the flood risk. This development along the rivers makes it more difficult to implement flood measures, such as compartmenting areas, creating water retention basins or floodways, viz. developing so–called *green rivers*.

In the Netherlands, it is typical to have continuous system of dikes that together with high ground areas and other hydraulic structures (gates, sluices, locks, pumping stations, etc.) create an enclosed area. These areas are called dike ring areas and currently there are 53 of them identified in the Netherlands (Figure 3.4). The largest dike ring surface area of 600 km^2 is in the Betuwe, in the so–called central river land. The Flood Defences Act of 1996 defined the safety standard for each dike ring area, "expressed as the average exceedance probability–per year–of the highest

Figure 3.4: Dike ring areas of the Netherlands.

water level, which the primary water defence must be capable of withstanding from the outside, while taking into account other factors which determine the water defensive capability" (Floris, 2005). For a dike ring, its weakest link defines its safety level or its strength. Representing the safety level of a dike ring by its weakest link is economically undesirable, as the failure probability of the entire dike ring is high, when only part of it has high probability of failure and needs particular attention.

Water retention and detention areas are seen as two potential solutions, both being applicable to particular situations. Flood water detention areas are useful for attenuating flood peaks, provided the storage volume is large compared to the flood peak and the detention occurs in time of peak discharge. The area is temporarily flooded and can be restored or be available for land use or other functions again later on. For flood reduction purposes, they are considered more efficient than retention areas that serve to store water, not allowing it to flow downstream. Water retention basins along the Rhine River are not seen as a feasible solution as they are considered to contribute only to the attenuation of low to medium peak flows, and not in the case of extreme or prolonged events. Detention areas far upstream in the Rhine basin are "not considered very effective in lowering extreme floods that endanger the downstream areas" (Hooijer et al., 2004). Clearly, some flood control solutions cannot be transferred to different places and are very much depending on the particular location and situation.

Dividing a dike ring area into compartments might significantly reduce the risk. During the Flood Risks and Safety (Floris) project, the re–estimation of flooding probabilities and consequences of flooding for 16 dike ring areas (out of 53 in total), together with in-situ examination of the dike rings was accomplished. The objective was to check the reliability of the dikes and the hydraulic structures, as well as to identify and reduce uncertainty. The risk was calculated for each dike ring area separately ignoring hydraulic interactions effects, that are not taken into account in the current design and safety assessment practice in the Netherlands despite the awareness of its importance. The damage in the province of South Holland varied approximately from €280 million to €37 billion (depending of the location of the breach) when the presence of embankments (highway, railroads, etc.) and natural terrain was considered, instead of €290 billion calculated assuming the flooding of the whole area without partitioning. For more details the reader is referred to the report by the Dutch Ministry of Transport, Public Works and Water Management (VenW, 2005).

Dutch are currently implementing the plan to give back some of the land to water. Creating room for the river means more storage for flood water by relocating the dikes further inland, lowering the floodplains, creating bypasses or so–called *green rivers*, creating detention areas along the river, etc. (see Figure 3.5). The *green river* definition is used since this solution is believed to give more priority to ecological benefits rather than only economic. It is a general belief that the 'Room for the River' policy is effective. However, its implementation at a large scale might be difficult especially when periodical flooding of inhabited areas is involved or when the water retention basins highly affect the economic development of the community.

The idea of amphibious floating houses and roads is currently on its way to implementation. The first floating village, expected to accommodate 20,000 people, is under construction at a semi aquatic city on IJburg near Amsterdam. This technology allows the houses to float on the water if the water levels rise during a flood event.

Vis et al. (2003) and Klijn et al. (2004) compared the controlling strategies of successively heightening and strengthening of dikes for coping with the Rhine River floods with two resilience strategies consisting of

1. Using detention areas of low economic value along the river and partitioning of a dike ring area into compartments with different flooding probabilities; or

2. Increasing river discharge capacity either by enlarging the floodplain area or by adding retention compartments of low economic value that have a high probability of flooding.

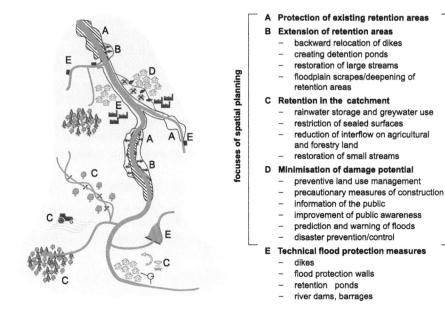

A **Protection of existing retention areas**

B **Extension of retention areas**
  – backward relocation of dikes
  – creating detention ponds
  – restoration of large streams
  – floodplain scrapes/deepening of
    retention areas

C **Retention in the catchment**
  – rainwater storage and greywater use
  – restriction of sealed surfaces
  – reduction of interflow on agricultural
    and forestry land
  – restoration of small streams

D **Minimisation of damage potential**
  – preventive land use management
  – precautionary measures of construction
  – information of the public
  – improvement of public awareness
  – prediction and warning of floods
  – disaster prevention/control

E **Technical flood protection measures**
  – dikes
  – flood protection walls
  – retention ponds
  – river dams, barrages

Figure 3.5: Illustration of typical location of flood risk management measures within the river basin (Hooijer et al., 2004).

The following criteria were assessed (Klijn et al., 2004):

1. Costs (investment and maintenance),

2. Flexibility (to adapt to changes in boundary conditions, normative views, etc.),

3. Resulting (expected) flood damage,

4. Economic impact,

5. Ecological impact,

6. Landscape qualities (scenery and cultural heritage).

The first (traditional) strategy was found economically favorable to implement being cheaper than the other two resilience strategies. While the first strategy involves the heightening of the dikes, the other two incorporate the construction of new dikes, adaptation of infrastructure, and financial compensation to the owners and inhabitants of areas where economic development will be affected as a result of increased allowed flooding frequency. The incremental costs of heightening existing dikes, is expected to be lower since the original costs of constructing dikes are not included in the overall estimation or comparison. Klijn et al. (2004) concluded that in long–term, the resilience strategies have fewer disadvantages than the flood controlling strategy of dike heightening. However, the implementation of the resilience

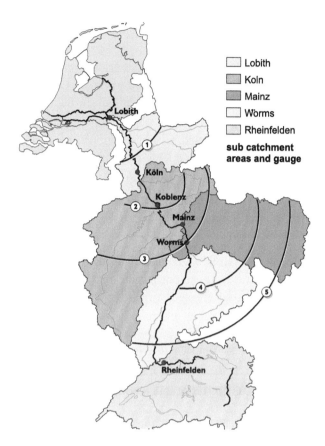

Figure 3.6: A map of the River Rhine basin with flood travel times (in days) to the Dutch–German border (Hooijer et al., 2004).

strategies requires large investments in the short–term whereas their revenues will become clear only after relatively long period of time.

Different types of flood risk management measures were considered during the IRMA –SPONGE project (Hooijer et al., 2004), focusing on flood prevention measures, development of methodologies to assess the impact of flood risk reduction measures and climate change scenarios. The objective was "to support the spatial planning process in establishing alternative strategies for an optimal realization of the hydraulic, economic and ecological functions of the Rhine and Meuse River Basins" (Wolters et al., 2001). Alternative resilience strategies have been elaborated and assessed for their hydraulic functioning and 'sustainability' in risk management criteria. The IRMA–SPONGE project (Hooijer et al., 2004) concluded that for the Rhine River:

1. The timing of flood peaks from tributaries to the main stream is highly complex.

2. The storage volume available in detention areas is utilized most effectively for peak shaving provided accurate timing of detention can be achieved during flood peaks and not during the earlier stages of floods.

3. The further upstream retention and detention areas are created, the less effective they are for reducing extreme floods downstream.

Downstream the Rhine River basin (Figure 3.6), the Flood Early Warning System (FEWS) is in operation. The water levels can currently be predicted with adequate accuracy 2 days ahead of a flood, and efforts are currently being made to lengthen this period up to 3 to 4 days.

As the flood risk increases, public awareness becomes a key factor for saving lives and it is important to have better participation of people in the decision making related to measures needed to decrease the risk. The population living in risk zones should be aware of the flood hazard maps. This is something that is not generally implemented in the Netherlands, but has been widely adapted in some other countries (e.g. disseminated by the Internet in UK).

3.3 Japan's experience in dealing with floods

Japan is a country where the occurrence of disasters is quite common. Earthquakes, floods, tidal waves, volcano eruptions and typhoons are part of every day life in Japan. Enclosed in all directions by the sea, it is vulnerable to storm surges, high waves, and tsunami on the coasts. Preserving the coastline from these hazards is crucially important while at the same time preserving various ecosystems. More than 50% of the population and more than 70% of the nation's assets are concentrated in the floodplains.

Storm surges by typhoons have caused catastrophic damages in Japanese coastal areas throughout the years. The typhoon Vera in 1959 caused Japan's greatest storm disaster accounting for nearly 5,000 casualties, and leaving some 1.5 million people homeless due to the enormous damage from wind, floods and landslides. If it were not for the construction of coastal dikes and improvement of weather forecast systems, the number of casualties from storm surge floods can continue to be high nowadays. However, the storm surge flood of 1999, the worst flood disaster since 1959, indicated that there is still a need to improve the measures against storm surges.

Rivers with their short length, steep slopes and narrow catchment areas are another source of flooding disaster. In response, engineers have taken disaster prevention and

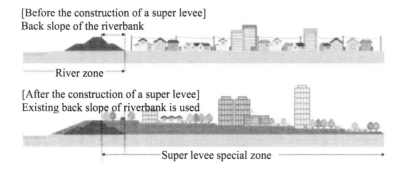

Figure 3.7: Example of super levee (www.rfc.or.jp).

Figure 3.8: Behaviour of conventional dike (above), and super levee (below) subject to overflow, seepage and earthquake shaking.

management measures. Flood control and management is implemented through a number of structural and non–structural measures. In addition to these mentioned in Section 3.1, artificial underground channels, stimulation of flood–proof building construction, installation of flood sensors that can assist in deciding on evacuation, and other options described in detail below are taken in Japan.

The underground floodways and regulating reservoirs are two solutions applied in Japan to deal effectively with urban flooding. The construction of super levees (Figure 3.7) is seen as a good solution to prevent flooding in urban areas where a potential dike break may have catastrophic consequences. In the battle against floods and earthquakes (Figure 3.8), they seem to offer stronger foundation and better protection. A super levee is considered resistant to overflow and seepage. Having a width of about thirty times their height, super levees have the strength to withstand severe flooding. Super levees provide usable land and space for dwellings and restores access to the riverfront.

The design tide level for coastal dikes is estimated based on the recorded values or the expected maximum water level creating uncertainty about any storm stronger than the designated storm. To deal with these exceptional events, early warning systems are in operation and evacuation plans are utilized. Urbanization near coastal

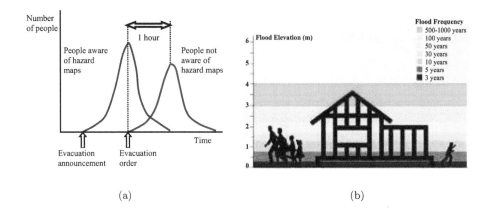

(a)   (b)

Figure 3.9: (a) Effectiveness of flood hazard map during torrential flood in Fukushima and (b) Expression of the safety degree against flood water levels (Merabtene et al., 2004).

areas and extensive expansion of mega cities in underground levels make people more vulnerable to potential flooding. For warning the population of the risk posed due to flood hazard, vulnerability maps are made available and are distributed to public.

Merabtene et al. (2004) confirms that warning the population using up to date information technology and preparatory work of flood hazard map distribution among them proves to be effective to mitigating flood damage and reduce the numbers of casualties. During the torrential rain of 1998 in Fukushima, people of Sukagawa city that were aware of flood hazard maps evacuated 1 hour earlier than those who were not (Merabtene et al., 2004). Figure 3.9a shows the difference in evacuation time among the people with and without knowledge about flood hazard maps.

A comprehensible Flood Risk Indicator has been developed in Japan to indicate the degree of safety/risk against flood damage through the means of a chart depicting the frequency of floods and inundation level versus the height of people and houses (see Figure 3.9b).

3.4 Modelling of propagation of flood caused by structural failure

To correctly estimate the consequences derived from a structural failure the modelling of flood propagation should be of high accuracy. Identification of the inundated areas, inundation depth, speed and duration, as well as the impact that flood water characteristics (salt, freshwater, contaminated water, etc.) can have on the inundated areas, are very important for decision making, emergency evacuation and early warning.

Most of the research on modelling dam break flood wave propagation has been focused on the movement of *clear* water in the downstream valley, excluding debris flows and sediment transport. *Clear* water dam break floods have been studied extensively experimentally, analytically and numerically.

The De Saint–Venant (1871) equations or shallow water equations are used for modelling dam break flood propagation. These equations consist of the mass and momentum conservation equations, and assume that the vertical velocities are much smaller than the horizontal velocities, which leads to hydrostatic pressure distribution in a channel cross section (Derivation of these equations can be found in Stoker (1957); Abbott (1979); Cunge et al. (1980); Chow (1959); Chaudhry (1993)).

At first, dam break flood analysis was assuming instantaneous and complete failure process rather than gradual failure of a dam. The sudden failure results into a highly unsteady flow, with a forward (positive) wave advancing in the channel downstream the dam and a back (negative) disturbance propagating into the still water upstream the dam (see Figure 3.10). Owing to their mathematical complexity, the analytical integration of de Saint–Venant equations in an unsteady flow situation can be obtained only for very few idealized situations. The first explicit solution to a dam break problem was given by Ritter (1892) who solved the de Saint–Venant equations analytically for the case of dam break flow in a horizontal and infinitely large rectangular channel neglecting the hydraulic resistance caused by stream bed friction and turbulence.

When the wave is considered to advance into still water of appreciable depth, the flow is approximated by a shock wave with jump conditions of mass, momentum, and energy conservation. The solution is deduced by evaluating the relation between hydrostatic forces and rate of change of momentum at the wave front. Many authors have carried out experimental investigations for the purpose of checking the validity of the analytical solution Schoklitsch (1926); Dressler (1954); U.S. Army Corps of Engineers (1960, 1961); Montuori (1965), etc.

The experiments of Schoklitsch (1917) indicated that the velocities for the forward wave may be as low as 40 percent of the result obtained by Ritter's solution, while having good agreement for the tip of the negative wave. Later, Dressler (1952) considered the wave propagation on a dry horizontal channel bed taking into consideration the effect of hydraulic resistance in the forward part of the flow. The hydraulic resistance caused by bed friction and turbulence in the shallow front region of the flow were considered dominant over the effect of the small slopes that were investigated later by Dressler (1958).

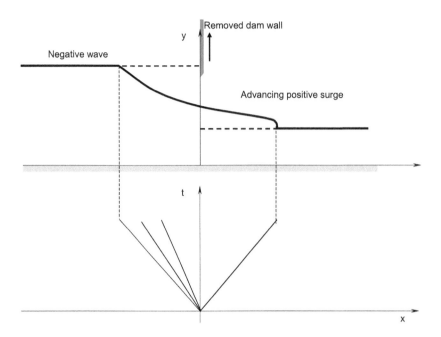

Figure 3.10: Dam break wave in a horizontal wet channel.

The numerical solution of the de Saint–Venant equations in natural streams was first achieved using the methods of characteristics. Later, Ré (1946) analyzed the dam break flood wave for a non–horizontal initially wet channel with a Chézy resistance coefficient using a finite–difference method applied to the characteristic equations. Different numerical schemes have been developed to solve the de Saint–Venant equations in 1 and 2 dimensional forms. Finite–difference (Abbott, 1979), finite–element methods (see e.g. Hervouet (2007)), and lately finite–volume (Toro, 1999) methods are generally preferred for their ability to reproduce discontinuous solutions.

Depending on the addressed problem and on the numerical scheme used to solve the shallow–water equations, conservative or non–conservative forms of the equations can be used. For different cross sections - arbitrary shape (natural shape), prismatic or rectangular - different forms of the shallow–water equations can be written. In Table 3.1 the conservative and non–conservative form of the shallow equations in arbitrary shape cross section are presented. For further information on modelling of dam break induced flows in complex topographies the reader is referred to Soares Frazão (2002); Chanson (2005).

During dam breaching, reservoir trapped sediment and dam material is moved away together with the water and is deposited in the downstream valley. According to Costa and Schuster (1988), historical floods from dam failure events have induced

Table 3.1: Shallow water equations in one dimension.

Represented forms	Continuity & momentum equation
Conservative form on arbitrary cross section, where: A = cross sectional area Q = discharge I_1 = first moment of the cross section I_2 = spatial variation of the cross section width S_0 = bottom slope S_f = friction slope q_l = lateral inflow per unit length	$\frac{\partial A}{\partial t} + \frac{\partial Q}{\partial x} = q_l$ $\frac{\partial Q}{\partial t} + g\frac{\partial I_1}{\partial x} + \frac{\partial}{\partial x}\left(\frac{Q^2}{A}\right) = gA\left(S_0 - S_f\right) + gI_2$
Non-conservative form on arbitrary cross section, where: U = velocity	$\frac{\partial h}{\partial t} + U\frac{\partial h}{\partial x} + h\frac{\partial U}{\partial x} = 0$ $\frac{\partial U}{\partial t} + U\frac{\partial U}{\partial x} + g\frac{\partial h}{\partial x} = g\left(S_0 - S_f\right)$

severe soil movements in various forms: debris flows, mud flows, floating debris and sediment–laden currents. Intense erosion and deposition is expected to highly affect the morphology of the valley that in turn affects the flow. Still, nowadays, the propagation of flood wave induced by a structural failure is commonly modelled ignoring any presence of debris and sediment. It is the complexity of the processes involved, that limits the modelling of floods induced by structural failure to *clear* water wave propagation. Despite the awareness of the importance that the proper modelling of flood water and sediment/debris interaction have on the hazard evaluation, mathematical or numerical modelling of this process is not yet satisfactory. For information on ongoing work on modelling of debris and sediment transport during structural failure see e.g. Leal et al. (2002); Nsom (2002); Spinewine and Zech (2007).

There are a number of commercial and non–commercial hydrodynamic packages available worldwide. Sobek 1D2D developed at Delft Hydraulics (www.sobek.nl), MIKE FLOOD at DHI (www.dhisoftware.com/mikeflood), Telemac at EDF-DRD (www.telemacsystem.com) are few of the most widely used commercial hydrodynamic computation products. There are also freely available software packages, mostly used for one dimensional modelling of the steady or unsteady flow.

In this thesis we use Sobek 1D2D, an integrated one– and two–dimensional numerical simulation package developed by WL | Delft Hydraulics for modelling the propagation of flood wave caused by a structural failure. Sobek 1D2D solves the shallow water equations using the so–called 'Delft Scheme' (Stelling et al., 1998; Stelling and Duinmeijer, 2003). The 1D schematization for flow through a river channel is combined with a 2D schematization for overland flow, bringing the model's behavior closer to the real physical behavior (see Figure 3.11). A control volume approach is used for the 1D calculation points of the channel in combination with the 2D grid cells. The flow in the 1D channel below the 2D grid bed level is treated as 1D flow, while the flow above the 2D grid level is treated as 2D flow within the area of the 2D grid cell.

Sobek 1D2D is capable of simulating the dynamics of overland flow over an initially dry land, as well as flooding and drying processes on every kind of geometry (Dhondia and Stelling, 2002). It can correctly simulate the transition between sub and supercritical flows and vice versa (Verwey, 2001). Stelling and Duinmeijer (2003) give a description of the grid scheme used in the two dimensional overland flow model of Sobek 1D2D, and present a comparison between numerical and experimental results of the dam break flood propagation experiments under dry and wet conditions. A GIS based tool is associated with Sobek 1D2D for data input and output processing.

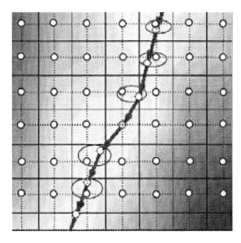

Figure 3.11: Schematization of the hydraulic model: Combined 1D/2D staggered grid.

3.5 Uncertainty associated with flood modelling

The processes involved in structural failure analysis are neither completely understood nor accurately described in mathematical terms. Uncertainty analysis is necessary for identifying sources of uncertainty and reducing it. Identifying the reducible and the irreducible uncertainties in a model is an important step in the overall process of uncertainty reduction. There are several types of uncertainties classified differently by various researchers (Funtowicz and Ravetz, 1990; Hoffman and Hammonds, 1994; Rowe, 1994; Krzysztofowicz, 2001; van der Sluijs, 1997). Vesely and Rasmuson (1984) classified uncertainty as:

1. Input or data uncertainty.

2. Modelling uncertainty, that is a result of incomplete understanding of the modelled phenomena, or numeral approximations used in mathematical representation.

3. Completeness uncertainty. The uncertainty associated with all omissions that occur due to lack of knowledge.

Tools for dealing with uncertainty are sensitivity analysis, error propagation equations, Monte Carlo analysis, expert elicitation, scenario analysis, Bayesian belief networks, etc. Saltelli et al. (2000) and Saltelli (2004) define the Sensitivity Analysis (SA) as the study of how uncertainty in the output of a model can be apportioned to different sources of uncertainty in the model input, and how a given model depends on the information fed into it. There are three types of sensitivity analysis: (i) Screening, (ii) Local Sensitivity Analysis, and (iii) Global Sensitivity Analysis. Local Sensitivity Analysis consist in varying one parameter at a time and observing its influence on the output, while Global Sensitivity Analysis consist in varying all parameters and evaluating their contribution to the variance of the output.

Janssen et al. (1994) defines uncertainty analysis as the study of the uncertain aspects of a model and their influence on the model output. Stochastic modelling is traditionally used to analyze uncertainty and uncertainty propagation associated with input data and model parameters that are represented by probability distribution functions rather than by a single value.

Monte Carlo Simulation and Latin Hyper Cube sampling are some of the methods for stochastic model calculations. Due to their time and calculation complexity, less time consuming sampling methods (best case, mean case, and worst case sampling methods) have been designed, that consequently provide less accurate information on the uncertainty in the outcome. The lack of information about the shape of the probability distribution functions of the values of the input data and model parameters leads to the idea of uncertainty of the uncertainty methods such as Monte Carlo or Latin Hyper Cube. Alcamo and Bartnicki (1987) argued that the uncertainty about the shape of probability distribution functions might be negligible.

In flood risk modelling, usually a sensitivity analysis is undertaken to detect the influence of the inputs on the output of the model. In this thesis we apply local sensitivity analysis to the models developed.

Chapter 4

Current Approaches to Dam/Dike Breach Modelling

There are lies, damned lies, and statistics.
Benjamin Disraeli

4.1 Background

One important step in dam break modelling is the accurate prediction of the breach outflow hydrograph. Significant effort has been made in developing models that accurately predict breach characteristics; still, many uncertainties related to breach modelling are still existent (Franca and Almeida, 2004; Morris, 2005; Zagonjolli et al., 2005; Zagonjolli and Mynett, 2005a). Due to the incomplete understanding of breach formation processes and hence the limited capabilities of mathematical description of dam breaching mechanisms, the presently available models rely on several assumptions as mentioned in Section 2.3.

In fact, no model is exclusively represented from first principles into mathematical equations and there always exist some degree of approximation and empiricism. In this thesis a distinction is made between empirical models and models created using data mining techniques, where a model is derived from the data itself, and physical models, where data are used to fit the equations based on physical principles.

During the IMPACT project, a European project investigating extreme flood processes and uncertainties*, several breach models were assessed using field and laboratory scale experimental data. Though these models offer improved capabilities and reliability in comparison with earlier models, their prediction accuracy is still far

---

*The reader is referred to the extra issue of Journal of Hydraulic Research, VOL 45, 2007.

from satisfactory. A 30% uncertainty band was identified for predicting peak outflows in the field and laboratory tests. When applied to the historical failure event of the Tous Dam in Spain, a 50% uncertainty band was observed. Poor prediction of the outflow hydrograph and unreliable prediction of the breach development in time was observed in the experimental results (Morris, 2005) suggesting that further research is needed to increase the capabilities of the existing physically based models. In this chapter we describe the present tools and methodologies used for breach modelling.

4.2 Experimental work

For better understanding the processes involved and to correctly describe the complex phenomena of embankment breaching, various field and laboratory experiments were carried out during the previous century. At first, the experimental work was focused on identifying the scale of flooding resulting from a sudden failure of the dam structure. This approach, though appropriate for concrete and arch dams that usually exhibit a failure within a relatively short duration of time - similar to sudden failure - did not provide insight into the breach modelling processes except for the flood wave movement in the downstream channel (Dressler, 1954). A thin plate representing the dam was located in a rectangular 'channel', and its removal simulated instantaneous failure of the dam.

During recent years, several field tests and laboratory experiments were carried out modelling gradual failure of dams. In the Netherlands, research regarding breach growth in dikes started with some field experiments on non cohesive sand dikes in 1989. Two fields tests were carried out at 'Het Zwin' located in Zeeuwsch-Vlaanderen, at the boundary between the Netherlands and Belgium. The experiment was focused on modelling sea sand dike breaching processes neglecting the effect of waves on the test results. The results of two other tests performed on 6th and 7th October 1994 are reported by Visser et al. (1996). Later, in 1996, large-scale field experiments were carried out for investigating the three dimensional aspects of breach growth in sand dikes (Visser, 1998).

Different failure modes and triggering conditions have been simulated in small scale laboratory tests and large scale field tests with the aim to better understand the breaching mechanisms and obtain data for calibration and validation of mathematical models. For further information on some of the experiments undertaken the reader is referred to Tinney and Hsu (1961); Tingsanchali and Hoai (1993); Coleman et al. (2002); Rozov (2003); Zhu et al. (2006). Here we focus on the latest experimental work carried out during the IMPACT project (IMPACT, 2004).

Several large scale field tests have been performed on non–cohesive and cohesive em-

Table 4.1: Information about five field tests undertaken in Norway.

No.	Description of the test
1	6m high clay and silt embankment (D50 = 0.01 mm) with 25% clay and less than 15% of sand to fail due to overtopping.
2	5m high non-cohesive embankment with D50 about 5mm (less than 5% fines) to fail due to overtopping.
3	6m rock fill embankment with moraine core to fail due to overtopping.
4	6m rock fill embankment with moraine core to fail due to piping.
5	4.5m moraine fill embankment to fail due to piping.

Table 4.2: Information about five laboratory tests undertaken at HR Wallingford (Morris, 2005).

Tests	Description of the test
9	Overtopping failure of homogeneous, non–cohesive embankments for various grain size, breach location and embankment geometry.
8	Overtopping failure of homogeneous, cohesive embankments with different material type and grading, compaction, water content and geometry.
5	Piping failure through homogeneous, cohesive embankments.

bankments failing from overtopping and piping failure modes. The five large scale field tests (see Figure 4.1) conducted in Norway are summarized in Table 4.1. The information about the twenty two laboratory tests conducted at a 1:10 scale at the HR Wallingford laboratories are summarized in Table 4.2.

These field and laboratory results lead to extensive validation of the numerical models being used and under development. In contrast to other experimental findings, during the field tests in Norway it was concluded that the breach shape is rectangular (vertical walls), not trapezoidal as often assumed previously. Morris (2005) suggest that the trapezoidal shapes might develop after the breaching is finalized, during the drying process of the embankment material. However, it is important to mention that in all laboratory experiments and field tests an initial rectangular breach channel was imposed as initial condition, which might have influenced the further development of the breach shape.

(a)

(b)

Figure 4.1: Development of the breach in the field tests in Norway for (a) piping failure and (b) overtopping failure (Vaskinn, 2003).

4.3   Empirical equations

An alternative, conservative way of estimating dam breach characteristics is the use of empirical predictor equations obtained from regression analysis of historical dam failure events. They are often used for providing ultimate breach extent information to physically based models (see e.g. Table 4.3), or as a comparative tool for the output obtained from physically based methods. However, their application to a particular dam should be used cautiously making sure that dam and reservoir characteristics are within the range of the dataset upon which the empirical equation is based.

Numerous empirical formulations have been developed for predicting dam breach characteristics and peak outflows based on hydraulic and geometrical characteristics of dams and reservoirs e.g. reservoir storage, depth and volume of water in the reservoir at the failure time, dam height, etc. In the following, some of the available empirical equations are described.

The U.S. Bureau of Reclamation (1988) suggested that for earthfill dams the ultimate width of a rectangular dam breach shape equals three times the initial water depth in the reservoir measured to the breach bottom elevation assumed to be at the stream bed elevation. This relationship was used as a guideline in the National Weather Service Simplified Dam Break Model (SMDBRK).

$$B_{avg} = 3h_w \tag{4.1}$$

where, B_{avg} is the ultimate average breach width and h_w is the water depth in the reservoir initiating the failure.

Hagen (1982) analysed 18 historical events of dam failure from overtopping. The product of volume V_w and depth h_w of water in the reservoir triggering the failure, hereafter called dam factor DF, was plotted versus peak discharge Q_p. As a result, Hagen provided an envelope equation:

$$Q_p = 1.205(V_w h_w)^{0.48} \tag{4.2}$$

The peak outflow discharge varied from 730 to some 83,000m^3/s while the dam factor $V_w h_w$ ranged from 2.9×10^6 to 4.8×10^{10}m^4.

MacDonald and Langridge-Monopolis (1984) analysed 42 dams: 30 earthfill and 12 non–earthfill dams (rockfill and other dams with protective concrete surface layers or core walls). The height of the dams varied from 6 to 93m. The product of water volume outflow V_{out} from the reservoir with the depth h_{w-b} was plotted versus the peak outflow for 18 earthfill dams and the following envelope and regression equations were obtained:

$$Q_p = 3.85(V_{out} h_w)^{0.41} \tag{4.3}$$

Table 4.3: Estimation of breach width and time of failure as a function of dam height H_d, crest length CL and different dam types.

Type of dam	Breach width	Formation Time
Earthfill & rockfill	$3H_d$	$\sqrt{H_d}$
Concrete gravity	$CL/3$	$\sqrt{H_d}$
Arch	CL	$H_d/50$

$$Q_p = 1.15(V_{out}h_w)^{0.41} \tag{4.4}$$

The volume V_{out} includes the reservoir storage at the failure time V_w plus the inflow associated with the rising limb of the inflow hydrograph.

Singh and Snorrason (1982, 1984) analysed some historical earthfill dam failure events due to overtopping. One finding of this analysis was the identification of a strong correlation between breach width and dam height. For most of the cases the breach width fell between the two bounding lines: $B = 2h_d$ and $B = 5h_d$. The maximum water depth above the dam crest triggering failure varied in a range from 0.15 to 0.61m, and the failure time from inception to completion of the breach, fell in the range between 0.25 to 1 hour. Singh and Snorrason (1982, 1984) analysed the hypothetical failure of eight earthfill dams in Illinois (USA) with a height h_d ranging from 4.4m to 28m. Six out of eight dams were simulated for failing from overtopping and two of them were simulated for failing from piping assuming a Probable Maximum Flood inflow in the reservoir. US Army Corps of Engineers HEC-1 and National Weather Service BREACH (Fread, 1988) models were used for predicting the peak outflow. In both models, the maximum failure time, maximum breach depth and width, and initial water depth above the dam crest (for overtopping failure) were predefined by the user equal to 0.5 hours, h_d, $4h_d$ and 0.15m, respectively. The obtained peak outflows were plotted versus the reservoir storage S and dam height h_d and the following regression equations were obtained:

$$Q_p = 1.78S^{0.47} \tag{4.5}$$

$$Q_p = 13.4h_d^{1.89} \tag{4.6}$$

Costa (1985) analysed 31 historical dam failure events. The height of the dams considered in the analysis varied from 1.8m to 83.8m, while the volume of the reservoir at the failure time ranged from $3.8\times10^3\mathrm{m}^3$ to $7.0\times10^8\mathrm{m}^3$. No distinction was made between different failure modes and dam types. The following envelope curves encompassing 29 data points were derived:

$$Q_p = 1.12V_w^{0.57} \tag{4.7}$$

$$Q_p = 2.63(V_wh_d)^{0.44} \tag{4.8}$$

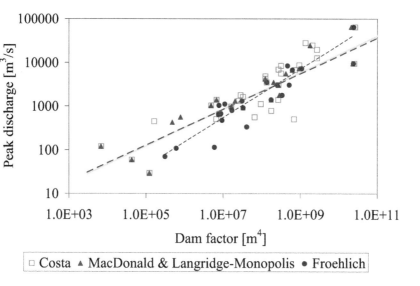

Figure 4.2: Dam factor vs. peak discharge for dams analysed by MacDonald and Langridge-Monopolis (1984), Costa (1985), and Froehlich (1987).

while, the regression relations are proposed as follows:

$$Q_p = 10.5 h_d^{1.87} \tag{4.9}$$

$$Q_p = 1.27 V_w^{0.48} \tag{4.10}$$

$$Q_p = 0.981 (V_w h_d)^{0.42} \tag{4.11}$$

There is no significant difference between the regression equations obtained by MacDonald and Langridge-Monopolis (1984) and Costa (1985), though the first authors include in the analysis only earthfill dams, while the second author includes earthfill and non–earthfill dams in his analysis.

Froehlich (1987) analysed 22 embankment dam failures with h_w ranging from 3.4m to 77.4m and V_w ranging from 0.1 to 310 million m^3. The regression equation obtained is expressed as:

$$Q_p = 0.607 (V_w^{0.295} h_w)^{1.24}. \tag{4.12}$$

In all analysis, more than 90% of the analysed dam failure events belong to dams in the USA. In Figure 4.2, peak outflow regression relations suggested by MacDonald and Langridge-Monopolis (1984); Costa (1985); Froehlich (1987) are presented along with measured peak outflows.

Broich (1998) analysed 39 embankment dam failures and obtained two regression equations which relate peak outflow to the depth and volume of water in the reservoir at the initial failure time in the following way:

$$Q_p = 255.859(V_w h_w)^{0.449} \qquad (4.13)$$

$$Q_p = 72.611(V_w h_w{}^4)^{0.256}, \qquad (4.14)$$

where V_w is in 10^6m^3.

Wahl (2004) carried out an uncertainty analysis of the empirical equations using a compiled database of 108 dam failure events. In the analysis no distinction was made between different failure modes (the same stands for Costa (1985) analysis). It is clear that for peak outflow prediction based on dam height and reservoir storage, the distinction between different failure modes is important. The water depth in the reservoir can be lower than the dam height in case of a piping failure mode, and higher than the dam height during an overtopping failure mode.

4.4   Physically based models

Cristofano (1965) as reported by Singh (1996) developed a dam breach model with the following characteristics: Breach Shape is considered trapezoidal during entire breach development; Bottom Width of the breach channel is assumed constant; Top Width is enlarged with the maximum width defined by the maximum side angle equal to the angle of repose of the compacted material of the dam. The initial channel is supposed to be parallel to the dam bottom and the (bottom) slope of the breach channel is assumed equal to the internal friction angle.

The flow through the channel is calculated using the broad–crested weir flow equation with a discharge coefficient ranging from 2.9 at the beginning of failure to 2.2 at peak flow

$$Q = CLH^{1.5}, \qquad (4.15)$$

where C is the discharge coefficient, H is the water depth above the breach bottom, L is the length of the breach channel, and Q is the discharge.

Sediment erosion is deduced by relating the force of the flowing water through the breach channel to the resistive shear strength of the soil particles acting on the bottom surface of the breach channel. The rate of erosion is expressed as a function of the rate of change of water flowing through the breach.

Harris and Wagner (1967) presented a simple model for the breaching of embankment dams due to piping and overtopping. The breach shape is assumed parabolic with top width of the parabolic shape B_{top} assumed to be 3.75 times the depth D (see Figure 4.3). The flow through the breach channel, assuming critical flow over a

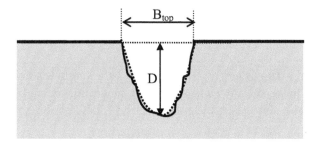

Figure 4.3: Parabolic breach shape.

broad–crested weir and accounting for the parabolic breach shape, is expressed in SI units as

$$Q = 5.54H^{2.5}. \tag{4.16}$$

Three modes of sediment erosion were adopted in the model: bed load, suspended load and saltation load. The Schoklitsch bed load formula (Schoklitsch, 1917) was modified to estimate the bed load transport along the breach channel. The rate of sediment transport was computed as

$$G = 86.7S^{1.5}Qd^{-0.5},$$

where G is the sediment transport rate, S is the hydraulic slope, and d is the soil grain diameter (expressed in English units). The suspended load had to be estimated based on the turbulence intensity and fall velocity. A constant sediment concentration is assumed and no slope stability or failure mechanism is considered.

Brown and Rogers (1977, 1981) developed the BRDAM model that simulates the breach erosion of earthfill dams due to piping or overtopping. The initial channel is created at the dam crest, when simulating overtopping failure. The parabolic breach shape of the notch is assumed constant during the whole process of dam breaching. Some 45° side slopes and a channel slope of 5° up to 20° is assumed, depending on the fill material of the dam. The outflow during overtopping failure is calculated by the broad–crested weir formula:

$$Q = \sqrt{g}B_{top}y_c^{1.5}. \tag{4.17}$$

The critical depth of the flow over the crest is taken to be 75% of the water depth above the breach bottom level. Like Harris and Wagner (1967), the Schoklitsch (1934) bed load formula is used to estimate the solid load which is transported along the breach channel bed assuming constant sediment concentration. The slope stability and lateral erosion mechanisms are not taken into account in the model.

Ponce and Tsivoglou (1981) presented a model for gradual failure of earthfill dams

due to overtopping. The initial location of the breach is assumed at a weak point
on the crest and downstream face. The top width of the breach opening is related
to the flow rate.

Lou (1981) developed a mathematical model that is similar to the Ponce and Tsivoglou
(1981) model. The breach grows in space and time until an equilibrium section is
attained. The most efficient stable section derived for describing the breach shape
is expressed as:

$$y = h_{max}cos(\frac{\pi}{B_{top}}x),$$

where y is the depth at a distance x, h_{max} is the maximum depth at the center,
and x is the distance from the center. De Saint–Venant equations are applied to
model the unsteady flow of water during breaching of an earth dam. The dam is
divided into n-reaches, with the discharge flowing from the reservoir routed downstream from the dam. During dam overtopping, the sediment discharge is computed
using three approaches: 1)Du Boy's equation (Du Boys, 1879) for bed load sediment
transport combined with Einstein's equation for suspended load sediment transport
to calculate the total sediment transport; 2) a sediment transport formula derived as
a function of failure duration, erodibility index, and water velocity, 3) Cristofano's
equation for computing the erosion rate. Du Boy's equation is written as:

$$q_{bl} = K\tau_0(\tau_0 - \tau_c), \qquad (4.18)$$

where q_{bl} is the rate of soil transported as bed load per unit width, $(\tau_0 - \tau_c)$ is the
excess shear stress, and K is a parameter dependent on the thickness of the moving
bed layer, velocity of the layer as well as critical shear stress τ_c.

Einstein's (1942) suspended load rate per unit width can be expressed as

$$q_{sl} = K_s \left[2.303log\left(\frac{30.2h}{d_{65}}\right)I_1 + I_2\right], \qquad (4.19)$$

where h is the flow depth, I_1 and I_2 are integral quantities depending upon the flow
depth, d_{65} is a representative grain size diameter, $K_s = 11.6C_a u_s a_{bl}$, a_{bl} is thickness
of the bed layer where the suspended load starts, $u_s = (gRS)^{0.5}$ and C_a is the
sediment concentration at a known distance from the bed.

Lou (1981) as described in Singh (1996) derived an expression for solid transport in
embankments by relating the kinetic energy to the erosion process, with the following
expression:

$$E_r = \frac{C_e}{12}t_d u^4,$$

where C_e is a constant, t_d is the duration of failure, u is the water velocity and M_s is
the mass of soil lost during erosion. Slope stability mechanisms are not considered,
while the lateral erosion stops once the peak flow is reached.

The National Weather Service model BREACH developed by Fread in 1984 and later revised in 1988 (Fread, 1988) simulates overtopping and piping failure modes for cohesive and non–cohesive embankment dams. The dam may consist of two materials with different properties in the inner core and outer zone of the dam or otherwise be homogeneous. Erosion initially occurs along the downstream face of the dam where a small rectangular–shaped rivulet is assumed to exist along the dam face. The sides of the breach channel are assumed to collapse when the depth of the breach reaches a critical value and then the breach is transformed into a trapezoidal shape channel. Maximum values of breach dimensions are defined by the user (dam height, maximum top and bottom breach width limited from the valley cross section). The flow through the breach channel is calculated by the broad–crested weir equation and the orifice flow relationship (for piping). The sediment transport rate is calculated by the Meyer-Peter and Müller (1948) sediment transport relation as modified by Smart (1984) for steep channels.

DEICH (Dam Erosion with Initial breach Characteristic) models are developed at the University of the Federal Armed Forces Munich (Broich, 1998). Depending on their calculation approach, they are named: DEICH_A (analytical model), DEICH_P (parametrical model), and DEICH_N1/N2 (1D/2D numerical model), respectively. These models calculate the overtopping failure of cohesive and non–cohesive embankment dams with or without covering core. Initially a channel or pipe is created at the dam crest. The breach develops parallel to the dam crest. A constant trapezoidal breach shape with side slopes of 45° and a constant relationship between lateral and bottom erosion is assumed. Breach discharge is calculated using the broad crested weir formula enhanced to take into account backwater effects. DEICH_A and DEICH_P calculate the total sediment load, while DEICH_N1/N2 distinguish between bed load and suspended load. DEICH_A uses Ponce–Tsivoglou's transport formula (Ponce and Tsivoglou, 1981), a combination of the Exner with the Meyer-Peter and Müller (1948) equation. Slope stability mechanisms are not considered in all models.

Several commercial hydrodynamic modelling packages have incorporated breach models. However, most of the time the final breach characteristics and rate of breach growth have to be predefined by the user who often bases his decision on expert judgment or empirical equations that relate breach parameters to dam and reservoir characteristics. Mike11, the Danish Hydraulic Institute's modelling package, includes a simple dam breach model. In this package the failure of the dam can initially take place from overtopping or as a piping failure. The breach will continue to develop until it has reached the breach geometry limit, which is defined by final bottom depth, width and the breach slope on each side of the breach. The development of the breach can take place in two different ways:

1. Time Dependent: The development of the dam breach is specified by the user in terms of breach level, width and slope as functions of time.

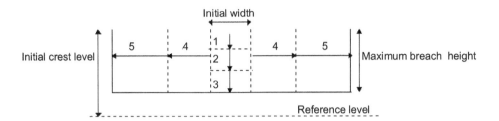

Figure 4.4: Breach development in Sobek 1D2D, first in vertical direction (step 1, 2, and 3) and then in horizontal direction (step 4 and 5).

2. Erosion Based: MIKE 11 calculates the breach development by using a sediment transport formula.

In Sobek 1D2D, WL | Delft Hydraulics' modelling package, a dam break can be modelled either in 1D or 2D, using the 1D Flow Dam Break Reach (mainly for flood defense structures along the water body) or the 2D-Breaking Dam Node (mainly for water retaining structures) respectively. The 1D Flow Dam Break Reach model (Verheij, 2002) can simulate breach development in clay or sandy dikes. The breach shape is considered rectangular during the entire breach development. The breach will first deepen in vertical direction only, until the maximum breach depth is reached, after which the breach grows in lateral direction only. Figure 4.4 presents the different breach development stages during the breach growth. The breach development can take place as either:

1. Irrespective of actual occurring flow conditions in the breach, or

2. Depending on the occurring hydraulic conditions and the parameters defined.

In the first case, the breaching development is user defined. The user must provide the initial breach width, and initial crest level of the dam, maximum breach depth, the starting time of the breaching process, the duration of the breaching process, the maximum breach–width or the maximum duration for reaching the maximum breach width. The discharge through the breach channel is calculated using standard weir formulas.

In the second case, the user must supply information about the lowest level that the breach will reach, the maximum flow velocity for which erosion does not occur, the starting time of the breaching process and the duration of the breaching process. In this case the increase in crest width as a function of time is not known beforehand, but depends on the occurring hydraulic conditions and the parameters defined.

For information about other breach models the reader is referred to Table 4.4 and

Singh (1996); Singh and Quiroga (1987); Loukola and Huokuna (1998). To further understand the mechanisms associated with embankment failure, we examine a database of the historical events. In the following sections, we present a methodology for estimating breach shape and breach outflow characteristics based on a data mining approach.

Table 4.4: Breach model characteristics.

Model	Geometry	Hydraulics	Sediment
Cristofano (1965)	Trapezoidal with side angle equal to angle of repose of the compacted material	Broad-crested weir	Developed empirical formula
Harris and Wagner (1967)	Parabolic with top width 3.75 times the depth	Broad-crested weir	Modified bed load Schoklitsch formula and suspended load formula based on turbulence and fall velocity
Brown and Rogers (1981) BRDAM model	Parabolic with side slope of 45	Broad-crested	Modified bed load, Schoklitsch formula
Ponce and Tsivoglow (1981)	Top width flow rate relation	de Saint-Venant	Exner with Meyer-Peter-Muller
Lou (1981)	Most effective stable section (Cosine curve shape)	de Saint-Venant	1. Du Boy & Einstein, 2. Lou, 3. Christofano.
Nogueira	Effective shear stress section (Cosine curve shape)	de Saint-Venant	Exner equation with Meyer-Peter-Muller
Fread (1988) NWS BREACH	Rectangular and trapezoidal	Broad crested weir orifice flow	Meyer-Peter-Muller modified by Smart
Singh & Quiroga (1987) BEED	Trapezoidal	Broad crested weir	Einstein-Brown
Loukola & Houkuna (1998) EDBREACH	Trapezoidal	Broad crested weir	Meyer-Peter-Muller
Broich (1998) DEICH_N1/N2	Diffusion approach	de Saint-Venant	Several transport eqs.

Chapter 5

Data Mining Techniques in Dam Breach Modelling

First get your facts, then you can distort them at your leisure.

Mark Twain

5.1 Introduction

While capabilities of computational modelling systems are continuously advancing, data from sensor measurements, satellites and computational operations are rapidly expanding. Hence, data mining techniques are becoming increasingly more popular nowadays and are being applied to different fields including water science and engineering (Minns and Hall, 1996; Price, 2000; Babovic et al., 2001; Mynett, 2002, 2004a,b, 2005; Solomatine, 2002).

Data mining techniques can be used to extract knowledge from large amounts of data. Despite the scarcity of reliable data related to historical dam failure events, it was hypothesized that present day data mining techniques might prove to be useful to some extent for predicting breach characteristics and peak outflow resulting from dam failure. Ultimate breach width and dam peak outflow might be predicted based on dam and reservoir characteristics at the failure time. In this chapter, we present a methodology for predicting dam failure characteristics and peak outflow using a range of data mining techniques, including linear and non–linear regression methods. The algorithms and underlying theory are introduced in the following section.

5.2   Main principles of data mining techniques

The engineering of intelligent machines has been a focus of scientific research for a long time. Already in 1950 Alan Turing (1950) proposed a test that would measure machine's capability to perform human–like conversation. It was John McCarthy (1956) who first introduced the term 'Artificial Intelligence (AI)' as the topic of the Dartmouth Conference in 1956, the first conference devoted to this subject, defining AI as "the science and engineering of making intelligent machines, especially intelligent computer programs". The ultimate goal of the artificial intelligence is to imitate human intelligence.

The field of machine learning is concerned with the question of how to construct computer programs that automatically improve with experience. It is a field based on computer science, psychology, neuroscience and engineering. In recent years many successful machine learning applications have been developed, ranging from data mining programs, to information filtering systems, all aiming to program computers in the way that they could "learn to improve automatically with experience" (Mitchell, 1997). There are different manners of machine learning, namely: supervised, unsupervised and reinforcement learning.

Supervised learning is a machine learning technique that creates a function based on training data of input–output pairs. The output of the function can be a continuous value (called regression), or a class label of the input object (called classification). The objective is to correctly predict the output given a new input after having seen a number of training examples.

In the unsupervised learning the model is fit to input data with no a priori output. The learner's task is to represent the inputs in a more efficient way, as clusters, categories or reduce the dimensions. Unsupervised learning is used for data compression, outlier detection, classification, etc.

Reinforcement learning (RL) is another learning technique much closer to supervised than unsupervised learning. It concerns learning from the consequences of action. The learner receives information that what it did is appropriate or not. Contrary to supervised learning where the learner is 'advised' on exactly what it should have done, reinforcement learning only says that the behavior was inappropriate and (usually) how inappropriate it was.

In this chapter we use the supervised learning techniques and algorithms, namely linear and non–linear techniques, for extracting information from the available dataset of historical dam failure events.

5.2.1 Linear regression

A linear model is written as

$$\mathbf{Y} = \mathbf{X}\mathbf{w} + \epsilon, \tag{5.1}$$

where $\mathbf{w}$ is the k–dimensional parameter vector, ϵ_i is the error vector usually assumed to be normally distributed with zero mean. The objective is to compute the parameter vector $\mathbf{w} = (w_1, w_2, ..., w_k)$ that best fits the data by the linear model. The linear model should be robust and not exhibit an unstable behavior that can greatly amplify even small changes in the dataset.

At the end of the 18th century, Gauss (1809) and Legendre (see Stigler, 1981) proposed the method of Least Squares (LS) regression that consists of minimizing the sum of the squared residuals with respect to the coefficient vector $\mathbf{w}$ as

$$\underset{\mathbf{w}}{\text{minimize}} \sum_{i=1}^{k} r_i^2, \tag{5.2}$$

where the residuals r_i are the differences between what is actually observed and what is estimated. However, this method is known to be highly influenced from the presence of outliers in the dataset. Thus other versions of this estimator were proposed such as replacing the squared residuals by the median of squared residuals (Rousseeuw, 1984; Rousseeuw and Leroy, 2003). The least absolute deviation method introduced in 1757 by Boscovich (Birkes and Dodge, 1993) (also known as least absolute value method) and least median of squares linear regression methods are considered better methods for regression and often recommended as robust or outlier resistant alternatives of LS, though they can exhibit instabilities for small changes in the data.

The least absolute value method, presented by Edgeworth (1887) is expressed as

$$\underset{\mathbf{w}}{\text{minimize}} \sum_{i=1}^{k} |r_i|. \tag{5.3}$$

In contrast to the LS method, it minimizes the sum of the absolute values of the residuals. Here, we use the Least Median of Squares (LMS) method (Rousseeuw, 1984) that is considered a less sensitive and more robust fitting technique compared with the simple linear regression method. It is also considered to be the most widely used robust estimator. It minimizes the median of the squared residuals with respect to the coefficient vector $\mathbf{w}$:

$$\underset{\mathbf{w}}{\text{minimize}} \, \underset{i}{\text{med}} \, r_i^2. \tag{5.4}$$

Thus the largest residuals in the sample (i.e. those whose absolute values are larger than the median) are ignored, making this technique more robust in the presence of outliers or noisy data. At least 50% of the data would need to be corrupt in order to skew the result, while for ordinary least squares a single corrupt data sample can already give the resulting regression line an excessively large slope.

5.2.2 Artificial Neural Networks (ANNs)

McCulloch and Pitts (1943) were the first to introduce the concept of how the brain could produce complex patterns by using basic cells called neurons that are connected to each other. McCulloch and Pitts (1943) presented an artificial neuron model with binary input and output and an activation threshold.

Artificial Neural Networks (ANNs) are envisioned to have similarities to the human brain functioning. As the latter are built of very complex webs of interconnected neurons, ANNs are built to form complex interconnected sets of units, each of which takes a number of real–valued inputs and produces a single real–valued output (Figure 5.1). They can be used for approximating discrete–valued, real–valued and vector–valued target functions.

The perceptron is a type of artificial neural network invented in 1957 by Rosenblatt (1958). The perceptron (Figure 5.2) takes a vector of real–valued inputs, calculates a linear combination of these inputs, and then outputs results based on some activation function. Using the McCulloch and Pitts (1943) threshold function, the output is 1 if the result is greater than the threshold value and 0 (or -1) otherwise. Therefore, given inputs x_1 through x_n, the output $o(x_1, ..., x_n)$ computed by the perceptron is

$$o(X_1, ..., X_n) = \begin{cases} 1 & \text{if} \quad\quad w_0 + w_1 x_1 + w_2 x_2 + ... + w_n x_n > 0 \\ 0 & \text{otherwise} \end{cases}, \quad (5.5)$$

where each w_i is a real–valued constant referred to hereafter as *weight*, determining the contribution of input x_i to the perceptron output. Teaching a perceptron network involves establishing values for the weights $w_0, ..., w_n$. Depending on the problem, one of the activation functions showed in Figure 5.3 can be used.

Several algorithms are known to solve this learning problem e.g. perceptron rule, delta rule, etc. (Mitchell, 1997). The perceptron rule finds a successful weight vector when the training examples are linearly separable. It revises the weight w_i associated with input x_i according to the rule:

$$w_i \leftarrow w_i + \Delta w_i, \quad (5.6)$$

where

$$\Delta w_i = \eta (t - o) X_i. \quad (5.7)$$

Here t is the target output for the current training example, o is the output generated by the perceptron, and η is a positive constant called the *learning rate*. The role of

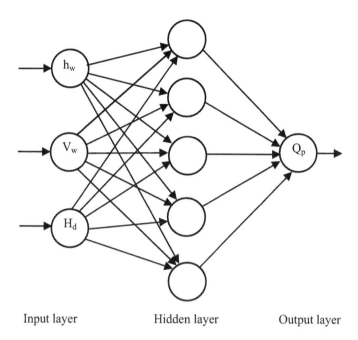

Input layer   Hidden layer   Output layer

Figure 5.1: Architecture of a three layer neural network with three inputs and one output.

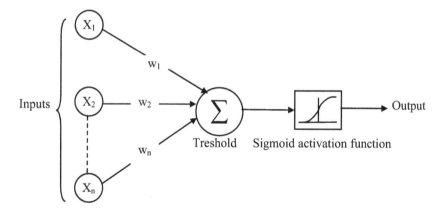

Figure 5.2: A perceptron.

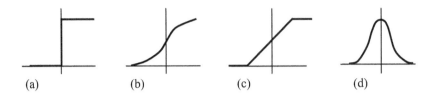

Figure 5.3: Commonly used activation functions: (a) Binary (threshold step function), (b) Sigmoid, (c) Piecewise Linear, and (d) Gaussian activation functions.

the *learning rate* is to control (lessen) the degree to which weights are changed every time step. The rule converges within a finite number of applications of the perceptron training rule to a weight vector that correctly classifies the training data, provided they are linearly separable and provided a sufficiently small η is used (Mitchell, 1997). Contrary to the perceptron rule that can fail to converge when the data are not linearly separable, the *delta rule* (or sometimes the LMS rule, Adaline rule, or Widrow-Hoff rule - after its inventors) converges towards a best–fit approximation to the target concept in that case. The *delta rule* uses the *gradient descent* to search the hypothesis space of possible weight vectors to find the weights that best fit the training data. This is done by specifying a measure for the training error of a hypothesis (weight vector) relative to the training examples such as:

$$E(\mathbf{w}) \equiv \frac{1}{2} \sum_{m \in M} (t_o - o_d), \tag{5.8}$$

where m is the set of training examples, t_o is the target output for training example m, and o is the output of the linear unit for training example m. To understand the gradient descent algorithm, we visualize the entire hypothesis space of possible weight vectors and their associated E values, as illustrated in Figure 5.4.

During the gradient descent search, a weight vector is determined that minimizes E by starting with an arbitrary initial weight vector, then repeatedly modifying the weights in small steps, during the learning phase. During each step, the weight vector is altered in the direction that produces the steepest descent along the error surface shown in Figure 5.4. The direction can be found by computing the derivative of E with respect to each component of the vector $\mathbf{w}$. This vector derivative is called the *gradient* of E with respect to $\mathbf{w}$, written $\Delta E\mathbf{w}$.

$$\Delta E(\mathbf{w}) \equiv \left[\frac{\partial E}{\partial w_0}, \frac{\partial E}{\partial w_1}, ..., \frac{\partial E}{\partial w_n} \right]. \tag{5.9}$$

The gradient specifies the direction that produces the steepest increase in E. The negative of the vector therefore gives the direction of steepest decrease. Single perceptrons can only express linear decision surfaces. In contrast, multilayer perceptron (MLP) networks trained with a *Backpropagation* algorithm are capable of capturing

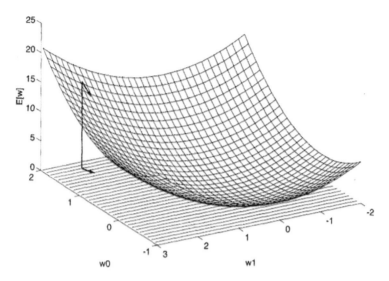

Figure 5.4: Hypothesis space of possible weight vectors and their associated error values E (Mitchell, 1997).

a variety of non–linear decision surfaces. The *Backpropagation* algorithm is the most commonly used ANN learning technique. Gradient descent provides the basis of the algorithm. It has been found to be appropriate for many problems, as summarized from Mitchell (1997), as follows:

- Instances are represented by many input attributes being correlated or independent of each other.

- The target function output might be discrete–valued, real–valued or a vector of several real–or discrete–valued attributes.

- Training examples might contain errors or noisy data.

- Training time is relatively long compare to other methods, depending on the number of weights in the network, the number of training examples considered, and the settings of various learning algorithm parameters.

- Despite the biological motivation of the neural networks, the learned neural networks are less easily communicated to humans and should be used when the readability of the results is not important.

Backpropagation algorithm uses a gradient approach descent to tune network parameters to best fit the training set of input–output pairs.

5.2.3   Instance Based Learning (IBL)

In contrast to above described learning methods which produce a generalization as soon as the data are introduced, there are the so–called instance based learning methods or 'lazy' learning methods that postpone the description of the target function until a new instance (query) is requested.

Each time a new query instance is encountered, its relationship to the previously stored examples is examined in order to assign a target function value for the new instance. These methods fit the training data only in a region around the location of the query instance instead of estimating the target function once for the entire instance space. Therefore, even a very complex target function can still be described, by constructing it from a collection of less complex local approximations using the instance based learning methods (Mitchell, 1997). These methods can construct a different approximation to the target function for each specific query instance requiring classification.

However, there are disadvantages associated with these kinds of methods such as: the cost of classifying new instances can be high since the computation takes place at classification time rather than when the training examples are first encountered. And since all attributes of the instances are considered when attempting to retrieve similar training examples from memory, the instances that are truly most 'similar' may well be a large distance apart.

The instance based learning methods include NEAREST NEIGHBOR, locally weighted regression methods, and case-based reasoning methods. The NEAREST NEIGHBOR method (IBk) chooses the closest points near the query point and uses its output. The weighted average model averages the outputs of neighbouring points, inversely weighted by their distance to the query point. The locally weighted learning regression model (LWL) weighs the outputs of neighbouring points according to their distance to the query point and then performs a local linear regression based on the weighted data.

The k-NEAREST NEIGHBOR method
The most basic instance-based method is the k–NEAREST NEIGHBOR algorithm that computes the classification of each new query instance as needed. It assumes all instances as points in the n–dimensional space $\Re^n$ and the nearest neighbours of an instance are defined in terms of the standard Euclidean distance.

Let an arbitrary instance x be described by the feature vector

$$\langle a_1(x), a_2(x), ..., a_n(x) \rangle, \tag{5.10}$$

where $a_r(x)$ denotes the value of the r–th attribute of instance x. Then the distance between two instances x_i and x_j is defined to be $d(x_i, x_j)$, where

$$d(x_i, x_j) \equiv \sqrt{\sum_{r=1}^{n}(a_r(x_i) - a_r(x_j))^2}. \tag{5.11}$$

In NEAREST NEIGHBOR learning the target function may be either real–valued or discrete–valued. When approximating real–valued (continuous valued) target functions, the k–NEAREST NEIGHBOR algorithm calculates the mean value of the k nearest training examples, where:

$$F(x_q) \leftarrow \frac{\sum_{r=1}^{k} f(x_i)}{k} \tag{5.12}$$

is a function that approximates the function $f(x)$ and that returns the mean value of f among the k training examples nearest to x_q.

For approximating a discrete–valued target function, the k–NEAREST NEIGHBOR algorithm denotes the k instances $x_1, ..., x_k$ from the training examples that are nearest to the query instance x_q and returns the most common value rather than the mean value of the k nearest training examples, where

$$F(x_q) \leftarrow \operatorname*{argmax}_{v \in V} \sqrt{\sum_{r=1}^{k} \delta(v, f(x_i))} \tag{5.13}$$

is a function that approximates the function $f(x)$ and that returns the most common value of f among the k training examples nearest to x_q. If k is chosen equal to 1 then the 1–NEAREST NEIGHBOR algorithm assigns to $F(x_q)$ the value $f(x_i)$ where x_i is the training instance nearest to x_q. For larger values of k, the algorithm assigns the most common value among the k nearest training examples.

The algorithm can better be explained through the graphical presentation of the Voronoi diagram (Figure 5.5). The instances are points in a two–dimensional space with the target function being boolean valued with positive and negative training examples shown by '+' and '-' respectively. The 1–NEAREST NEIGHBOR algorithm classifies the query instance x_q, as positive, whereas the 5–NEAREST NEIGHBOR algorithm classifies it as a negative example. The diagram shows the shape of the decision surface induced by 1–NEAREST NEIGHBOR over entire instance space.

The distance between instances is calculated based on all attributes of the instance. As such it is important to select the most important defining attributes. Large number of non important (irrelevant) attributes might influence the distance between neighbours and as a result lead to misclassification. The k–NEAREST NEIGHBOR algorithm is robust to noisy training data and quite effective when it is provided a

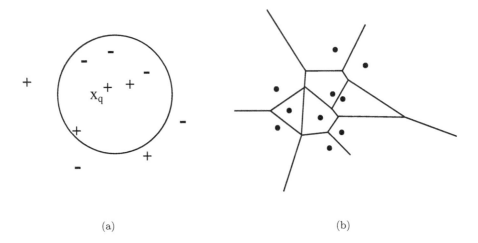

(a)   (b)

Figure 5.5: (a) Example of NEAREST NEIGHBOR and (b) Voronoi diagram.

sufficiently large set of training data.

Distance-weighted **NEAREST NEIGHBOR**

This algorithm weighs the contribution of each of the k neighbors according to their distance to the query point x_q giving greater weight to closer neighbors. Thus the algorithm formulation for approximating a real–valued target function changes to

$$F(x_q) \leftarrow \frac{\sum_{r=1}^{k} w_i f(x_i)}{\sum_{r=1}^{k} w_i}. \tag{5.14}$$

In a similar way, the algorithm formulation for approximating a discrete–valued target function changes to

$$F(x_q) \leftarrow \underset{v \in V}{\mathrm{argmax}} \sum_{r=1}^{k} w_i \delta(v, f(x_i)), \tag{5.15}$$

where

$$w_i \equiv \frac{1}{d(x_q, x_i)^2}. \tag{5.16}$$

Considering all training instances while approximating the function without the cost of misleading is possible with distance weighting as the very distant examples will have very little effect on $F(x_q)$. The only disadvantage of considering all examples is the 'longer' processing time. Furthermore, by taking the weighted average, the impact of noisy training examples will be smoothed.

Locally weighted regression

The locally weighted regression method constructs an explicit approximation to f over a local region or only on data near the query point x_q. The contribution of each training example is weighted by its distance from the query point. The distance-weighted training examples are used to form a local approximation to f. The target function in the neighborhood surrounding the instance x_q can be approximated using a linear function, a quadratic function, a multilayer neural network, or some other functional form. Here, let us consider the case when the target function f is approximated near x_q using a linear function of the form

$$F(x) = w_0 + w_1 a_1(x) + ... + w_n a_n(x). \tag{5.17}$$

As before, $a_i(x)$ denotes the value of the ith attribute of the instance x. Fitting the local training examples is done through an error criterion $E(x_q)$, expressed in three different manners.

1. Minimize the squared error over just the k nearest neighbors:

$$E_1(x_q) \equiv \frac{1}{2} \sum_{x \in k_{nrs}} (f(x) - F(x))^2. \tag{5.18}$$

2. Minimize the squared error over the entire set of training instances M, while weighing the error of each training example by some decreasing function N of its distance from x_q:

$$E_2(x_q) \equiv \frac{1}{2} \sum_{x \in M} (f(x) - F(x))^2 N(d(x_q, x)). \tag{5.19}$$

3. Combine 1 ands 2:

$$E_3(x_q) \equiv \frac{1}{2} \sum_{x \in k_{nrs}} (f(x) - F(x))^2 N(d(x_q, x)). \tag{5.20}$$

Second criterion allows every training example to influence the classification of x_q, but at the same time increases the computation time. The third criterion minimizes the squared error only over the nearest neighbors and thus the computation cost depends only on the number of k neighbors rather than the number of all training examples.

5.3   Application

5.3.1   Data preparation and model evaluation

During this research, we use the Waikato Environment for Knowledge Analysis (WEKA) modelling package (Witten and Frank, 2000). It is a comprehensive suite of Java class libraries that implement many state–of–the–art machine learning and data mining algorithms. Tools are provided for preprocessing data, creating models using a variety of learning schemes, and analyzing the resulting classifiers and their performance. The techniques described in the previous section are applied to the relatively small database of 108 historical dam failure events compiled by Wahl (2004). For each dam failure event the information consists of:

1. Dam characteristics before failure occurred: dam type, height and length, width at crest and bottom, upstream and downstream slope;

2. Reservoir characteristics: area and storage capacity;

3. Failure characteristics: failure mode, volume and depth of water at initial failure time, final breach characteristics (breach shape, top, bottom or average breach width and breach height, breach side slope factor), material removed from the dam, breach formation time and peak outflow.

Data are often incomplete, lacking attribute values, noisy, containing errors and outliers, or containing discrepancies in their information. Dataset does not provide enough information for prediction of time needed to initiate a breach and the rate of breach formation. Still, it might be useful for the prediction of breach characteristics and peak outflow.

We apply data preprocessing for transforming the raw data into a format that will be more easily and effectively used by data mining techniques. The first step in this analysis consists of visual inspection of the data, dealing with missing and inaccurate values and calculating statistical properties. Domain knowledge is applied for filling in missing values and resolving inconsistencies as shown in Table 5.1. In the first row for each of the dams some information is missing or is considered inaccurate. The given water depth and volume in the reservoir at the failure time for Swift Montana dam that failed from overtopping are substituted with values higher than the dam height and reservoir storage respectively. For Kendall dam the missing values of the water volume and depth in the reservoir at the time of the failure are assumed to be larger than maximum dam specification, for the reason that it failed due to overtopping.

The correlation between available attributes is calculated by dividing their covariance by the product of their standard deviations. We measure the accuracy with which one variable can be predicted from knowledge of another one by correlation coefficient. Dam height, reservoir storage and the water depth behind the dam at the initial failure time are used for the prediction of the breach characteristics and peak outflow. However, due to the absence of adequate information for all the dams, we use only a subset of the original database for experiments.

The problem of dealing with small datasets is that there is not enough data for train-

Table 5.1: Examples of missing values and inconsistencies.

Dam name (location) Failure mode	h_d	h_w	S	V_w
Swift Montana (USA)	57.61	47.85	3.7E+07	3.7E+07
Overtopping		≥ 57.61		>3.7E+07
Kendall Lake (USA)	5.49	?	7.28E+05	?
Overtopping		≥ 5.49	7.28E+05	$\geq 7.28E + 05$

ing and testing the model. Thus, we consider a 10-fold cross validation procedure in this study. In the cross validation method, the result does not depend on how the dataset is divided: every instance is used only once in testing. In this case, the dataset is randomly reordered and then split into ten folds of (approximately) equal size. During each iteration one fold is used for testing and the rest for training the model. The test results are collected and averaged over all folds. The Root Mean Square Error (RMSE) is used to measure the model performance:

$$RMSE = \sqrt{\frac{1}{n}\sum_{i=1}^{n}(y_i - \hat{y}_i)^2}, \tag{5.21}$$

where y_i and $\hat{y}_i$ represent the observed and computed values for the calculated attribute, whether it is average breach width or peak outflow. Lower RMSE values guarantee better performance of the model.

5.3.2   Predicting breach width based on water depth behind the dam

The first problem addressed in this section is the prediction of average breach width based on the water depth behind the dam at the initial failure time. The database (Wahl, 1998) provides a 'training' sample of previously observed cases in which both water depth in the reservoir at initial failure time and the average breach width have been recorded. Here, we construct a data–driven model for predicting likely average breach width values for future dam failure events, when only the water depth in the reservoir at the initial failure time is known.

For this analysis, only the overtopping failure mode is considered. In the dataset, most of the 23 dams belong to earthfill, mixed earthfill and rockfill types with or without revetment on the dam slopes. For 20% of the dams no information is available on their material type. Table 5.2 shows the performance of different algorithms applied to the available dataset. We apply the LMS (Section 5.2.1) algorithm to the available data using the cross validation procedure and use the equation obtained from LMS on the dataset (23 dams). The RMSE obtained is about

Table 5.2: Prediction of average breach width for 23 dams.

Method	RMSE
LMS1: $B_{avg} = 4.6h_w - 5.7$	15.0
USBR: $B_{avg} = 3h_w$	22.5

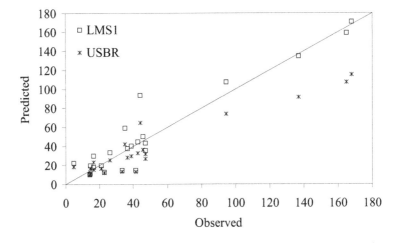

Figure 5.6: Average breach width prediction based on the water depth in the reservoir.

30% lower than the error when using the USBR equation on the same dataset, thus making former more applicable for predicting the average breach width in case of dam failure from overtopping. Figure 5.6 shows the observed and predicted average breach width computed using the LMS algorithm and the USBR equation. As can be seen from Figure 5.6, the USBR prediction tends to underestimate the average breach width, while LMS prediction results in a correct or overestimated average breach width.

Since in the dataset, 84% of the dams failing from overtopping belong to the low and medium dam categories, another experiment is performed for predicting breach width excluding the high dams from the dataset. It can be seen from Table 5.3 and Figure 5.7 that the LMS model performance increases by 30%, while the USBR equation performs 40% better than in the previous experiment. Both LWL and LMS algorithms perform 20% and 10% better, compared to the USBR equation.

From both experiments different equations (LMS1 and LMS2) are obtained, the second one demonstrating better performance on the dataset used for this analysis. It can be seen that for the prediction of breach width for non–high earthfill dams

Table 5.3: Prediction of average breach width for 18 low and medium dams.

Method	RMSE
LWL using inverse–distance weighting kernels	9.6
LMS2: $B_{avg} = 3.8h_w - 1$	10.9
USBR: $B_{avg} = 3h_w$	12

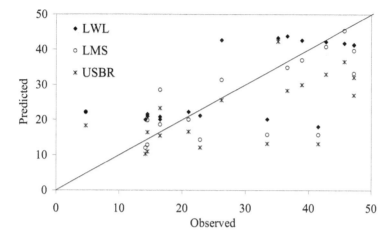

Figure 5.7: Average breach width prediction based on the water depth behind the dam at initial failure time for low and medium dams.

the LMS2 and USBR equations are performing almost the same with very small difference in RMSE.

5.3.3 Predicting peak outflow

In this section we address the issue of estimating the peak outflow based on dam height and reservoir storage. A total of 27 dams are included in the analysis for estimating peak outflow as a function of dam height. No distinction is made between different dam materials and failure modes. We consider this assumption as valid due to small dataset size, though the breach outflow hydrograph and consequently the peak outflow is likely to be affected by geotechnical characteristics of the embankment.

As can be seen from Table 5.4, models created with IBk and MLP perform comparably better than the Singh & Snorrason equation in case of peak outflow prediction based on dam height only. Two neighbours are used for classification in case of IBk application, while no changes are made in the default settings for the MLP network in WEKA. The decision to use default parameters is motivated by the scarcity of

Table 5.4: Prediction of peak outflow $Q_p = f(h_d)$.

Method	RMSE
IBk–2 neighbours	1000
MLP	1367
Singh and Snorrason (1982)-Eq. 4.6	2342

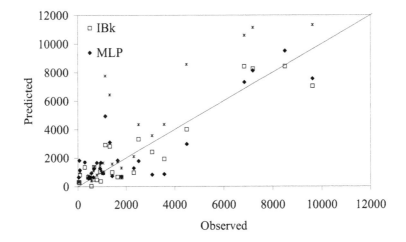

Figure 5.8: Peak outflow prediction based on dam height.

data and therefore the difficulty of having a separate validation set for parameter estimation. Though IBk RMSE estimate is high, it is 57% smaller compared to the one obtained from the Singh & Snorrason's empirical equation. The performance of IBk and MLP model and the Singh & Snorrason equation is presented in Figure 5.8.

Due to missing data, a smaller dataset of 22 dams is used for predicting peak out-flow based on reservoir storage and dam height. The performance of the developed data mining models is compared with the available empirical formulations of Singh & Snorrason, Hagen and Costa (Table 5.5).

Using the same data mining algorithms the results obtained by including the reser-voir storage in the analysis are compared to the empirical equations of Costa (Figure 5.9). Here, better model performance is observed using MLP rather than with IBk algorithm. Higher errors are observed when Hagen and Costa equations are applied to the whole dataset. However, better results can be expected if the distinction between failure types is taken into consideration, which in turn implies the necessity of more data.

Table 5.5: Prediction of peak outflow $Q_p = f(h_d, S)$.

Method	RMSE
IBk–2 neighbours	1200
MLP	1135
Hagen, Eq. 4.2	8400
Costa, Eq. 4.11	1300

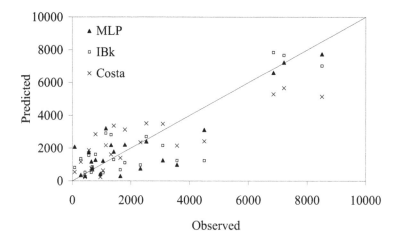

Figure 5.9: Peak outflow prediction based on dam height and reservoir storage.

5.4 Conclusions

Despite the shortage of documented data from historic dam failure events, in this chapter we demonstrate that it is possible to improve currently available empirical relations and prediction capabilities for breach characteristics by applying data mining techniques. Encouraging results are obtained for predicting the average breach width based on the water depth behind the dam at the initial failure time. The applied data mining algorithms, LMS and MLP, perform better than the available empirical equation of USBR (1988).

A 30% improved performance of the LMS and MLP models is observed when the complete dataset is used. Since 84% of the dams in the dataset belong to low and medium category, another experiment is carried out excluding the instances belonging to high dams. An improvement of the LMS and MLP model performance (9.6 and 10.9 RMSE respectively) and USBR equation (12.0 RMSE) is observed, leading to the conclusion that the data mining models seem to perform slightly better in experiments for predicting average breach width for low and medium earthfill dams failing from overtopping.

Furthermore, the prediction of peak outflow based on dam height and reservoir storage provides better results compared to the prediction of peak outflow based on dam height only. The results obtained by using IBk and MLP algorithms are better than the ones obtained by applying the empirical equations developed by Singh & Snorrason, that predict peak outflow based on dam height only, and of Hagen and Costa, whose equations use dam height, depth and volume of water in the reservoir for peak outflow prediction. Clearly, the relatively small number of recorded dam failure events contained in the database, limits the capabilities of data mining techniques. However, with increasing amount of information about historic (or future) dam failure events, better results might be expected.

Chapter 6

BREADA model: Breach Model for Earthfill Dams

One certainly cannot predict future events exactly if one cannot even measure the present state of the universe precisely.

Stephen Hawking

6.1 Development of a dam breach model

In this chapter we propose a method for modelling the overtopping failure of earthfill dams. In dike failure literature a distinction is made between overtopping and overflowing failure. The first is identified as the failure initialized from the waves (e.g. wind waves) flowing over the structure's crest for considerable amount of time, though the water depth in the water body is (still) lower than or equal to the structure's height. The overflowing failure is caused due to continuously rising water level (in the water body) above the structure's crest in absence of waves. The overflowing definition is almost never used in (embankment) dam failure literature. Instead, the term overtopping is generally used to describe the flow over the dam's crest irrespective of the waves' presence. Here, we also use the overtopping term though we do not consider waves in our model, but only the free flow over the dam, initiating a failure. The overtopping failure mode is known to be the most common failure mode (Costa, 1985; Singh, 1996) for embankment dams caused as a result of extreme inflows, malfunction of any hydraulic structure designated to release part of the hydraulic load, etc.

Controversial conclusions are drawn by different researchers (Johnson and Illes, 1976; MacDonald and Langridge-Monopolis, 1984; Morris, 2005) related to the initial, progressive, and ultimate breach shape, since proper monitoring of a real dam failure event is difficult to achieve so far. During laboratory and field experiments the initial breach shape is commonly predefined and thus influences in some way its further development (see Section 2.3.1). In the model BReaching of the EArthfill DAm (BREADA) it is possible to choose between two breaching developments. One formulation assumes a trapezoidal breach shape to be initially formed at the crest of the dam, which progresses to the dam bottom elevation (usually assumed at streambed elevation at the dam toe). Once the dam bottom is reached, the breaching develops in a lateral direction only. In the second formulation, a triangular breach channel is initially formed and progresses to the dam bottom (see Figure 6.1). Once the dam bottom is reached, the breaching develops in lateral direction having a trapezoidal shape. The latter is considered by several authors (Johnson and Illes, 1976; MacDonald and Langridge-Monopolis, 1984; Macchione and Rino, 1989) to be a common type of breach evolution during historical dam failure events.

The initial breach location, though an important component in breach modelling, is usually difficult (if not impossible) to predict beforehand, since many factors can affect any potential initial location, such as:

- Bad compaction quality of the dam material during construction, or inappropriate material used in dam construction.

- Presence of internal structures and weak points inside the dam's body.

- Damage to the dam structure that can occur during operation life, possibly at the slopes or at the crest.

The breaching of a dam usually initiates at the weakest point on the dam crest or downstream slope when overtopping occurs. In our model, the initial breaching is assumed to occur at the dam crest in the middle of the dam length (unless specified otherwise) and further breaching develops parallel to the dam crest (the implemented breach development mode is described in Section 2.3.2).

For routing of the flow through the breach, the reservoir routing principle is applied:

$$\frac{dV}{dt} = Q_{in} - Q_b - Q_c - Q_{outlet}, \tag{6.1}$$

where Q_{in}, Q_b, Q_c and Q_{outlet} are the inflow into the reservoir, the flow through the breach, the flow over the crest, and the flow through the outlet, respectively and dV/dt represents the change of reservoir volume in time.

Significant uncertainty is associated with experimental results obtained from laboratory and field experiments that determine the dominant sediment transport mechanisms during dam breaching (Section 2.3.3). For non–cohesive embankment dams, the erosion in a breach channel is considered to be similar to a morphological bed evolution (Broich, 2003). While in rare cases the erosion of dam material is assumed to be similar to *suspended load* transport, most of the time it is simulated as a *bed*

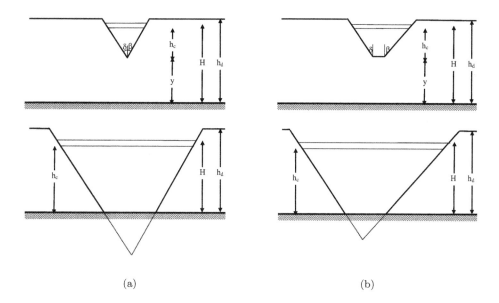

(a)   (b)

Figure 6.1: Breach development in both formulations: (a) triangular shape to the bottom of the dam and afterwards trapezoidal shape and (b) trapezoidal shape with vertical and lateral erosion to the bottom of the dam and afterwards lateral erosion only.

load transport (see Section 4.4 and Table 4.4). In the BREADA model, the erosion of the embankment material is calculated using an empirical formula presented by Meyer-Peter and Müller (1948).

Meyer-Peter and Müller (1948) performed the experiments on uniform bed material as well as with mixtures of particles with diameter ranging from 0.4 to 29mm, channel slope from 0.0004 to 0.02, water depth of 0.1 to 1.2m and specific weight of particles in range from 0.25 to 3.2N/m^3. The bed load transport rate [m^2/s] empirical equation obtained was written as

$$q_{bl} = \phi_b \left[(s-1)\,g\right]^{1/2} d_m^{3/2},\tag{6.2}$$

where s is the relative density ρ_s/ρ, ρ_s is soil density, d_m is the mean particle diameter [m] considered 1.1 to 1.3 times the d_{50} for nearly uniform material and equal to the d_{50} for mixture of materials (van Rijn, 1993), and ϕ_b is the bed load transport rate expressed in dimensionless form as

$$\phi_b = 8 \left(\mu\theta - \theta_{cr}\right)^{3/2},\tag{6.3}$$

where θ_{cr} is generally interpreted as the critical mobility parameter that for mixtures of sediments on stream beds it is usually taken equal to 0.047. During a dam breach erosion, θ_{cr} might be considered negligible compared to θ that is the dimensionless particle mobility parameter estimated as

$$\theta = \frac{RI}{(s-1)\,d_m}.$$ (6.4)

A bedform factor or efficiency factor μ is calculated with

$$\mu = \left(\frac{k}{k'}\right)^{3/2},$$ (6.5)

where k' is the grain–related Strickler coefficient and k is the overall Strickler coefficient. Yalin (1972) emphasized that if the plane is flat and the flow is two dimensional then the ratio k/k' becomes 1.

Substituting Eq. 6.3, 6.4, 6.5, into Eq. 6.2 we obtain

$$q_{bl} = k_{er}\,(\gamma RS)^{3/2},$$ (6.6)

where k_{er} is an erodibility coefficient equal to

$$k_{er} = \frac{8}{(\gamma_s - \gamma)\,g^{1/2}},$$

R is the hydraulic radius, γ is the water specific weight, γ_s is the soil specific weight, and S is the energy slope that can be calculated with Strickler's equation

$$S = \frac{u^2}{k^2\,R^{4/3}},$$

where u is the mean flow velocity. Assuming that the solid discharge entering our domain from the reservoir is null and considering a space step $\Delta x = 1$m, we can write the following mass conservation:

$$\frac{dA_b}{dt} = \frac{dq_{bl}c}{dx}$$ (6.7)

as

$$\frac{dA_b}{dt} = q_{bl}c,$$ (6.8)

where c is the wetted perimeter.

We assume that the initial conditions - water depth in the reservoir, initial breach depth and the inflow hydrograph - are known. For each time step, the flow over the crest, through the spillway and the outlet is calculated based on reservoir elevation. The difference between inflow and outflow is then used to calculate the change in reservoir storage and the new reservoir elevation. The reservoir stage–volume relationship is provided to the model through a mathematical expression. Based on this relationship the reservoir water level is extracted for every estimated reservoir storage at each time step. Other flow characteristics specific for different breach shapes are presented in the following.

6.1.1 First formulation of dam breaching development

The breach shape is assumed to be trapezoidal. At first the breaching of the dam progresses in vertical and lateral direction until the bottom of the dam is reached. The breach opening area is calculated as

$$A_b = B\left(H_d - y\right) + \frac{\left(H_d - y\right)^2}{2}\left(\tan\beta + \tan\delta\right), \tag{6.9}$$

where B is the bottom breach width. The development of breach cross section in time can be written:

$$\frac{dA_b}{dt} = \frac{dA_b}{dy}\frac{dy}{dt} = -\left[B + \left(H_d - y\right)\left(tan\beta + tan\delta\right)\right]\frac{dy}{dt}. \tag{6.10}$$

Flow in the breach channel is assumed a critical flow and the outflow discharge is calculated as

$$Q_b = \left(\frac{g}{B + h_c\left(\tan\beta + \tan\delta\right)}\right)^{1/2}\left[\frac{\left(2B + h_c\left(\tan\beta + \tan\delta\right)\right)h_c}{2}\right]^{3/2}, \tag{6.11}$$

where h_c is expressed as:

$$h_c = \frac{-3B + 2\left(H - y\right)\left(\tan\beta + \tan\delta\right) + R^{1/2}}{5\left(\tan\beta + \tan\delta\right)}, \tag{6.12}$$

and

$$R = 9B^2 + 8B\left(H - y\right)\left(\tan\beta + \tan\delta\right) + 4\left(H - y\right)^2\left(\tan\beta + \tan\delta\right).$$

Combining the Eqs. 6.6, 6.8 and 6.10, the breach bottom elevation can be estimated every time step with following equation

$$\frac{dy}{dt} = -\frac{k_{er}}{k^3}\frac{\left(\gamma g\right)^{3/2}}{2}\frac{\left(2B + h_c\left(\tan\beta + \tan\delta\right)\right)h_c}{B + \left(H_d - y\right)\left(\tan\beta + \tan\delta\right)}\left(\frac{B + h_c\left(\frac{1}{\cos\beta} + \frac{1}{\cos\delta}\right)}{B + h_c\left(\tan\beta + \tan\delta\right)}\right)^{3/2}. \tag{6.13}$$

For the development of the breach only in lateral direction (once the dam bottom is reached) the model equations for the breach characteristics are given as follows. The breach area is estimated as

$$A_b = \left(B + \left(\frac{H_d}{2} - y\right)\left(\tan\beta + \tan\delta\right)\right)H_d, \tag{6.14}$$

and the development of the breach only in lateral direction is expressed as

$$\frac{dy}{dt} = -\frac{k_{er}}{k^3}(\gamma g)^{3/2}\frac{h_c^2\left[2B - (2y - h_c)(\tan\beta + \tan\delta)\right]}{2H_d}\frac{\cos\beta + \cos\delta}{\sin(\beta + \delta)}$$

$$\times \frac{\left[B - y(\tan\beta + \tan\delta) + h_c\left(\frac{1}{\cos\beta} + \frac{1}{\cos\delta}\right)\right]^{1/2}}{\left[B + (h_c - y)(\tan\beta + \tan\delta)\right]^{3/2}}. \tag{6.15}$$

The discharge through the breach opening is estimated to be

$$Q = \sqrt{\frac{g}{2}}\frac{\left[h_c\left(2B - (2y - h_c)(\tan\beta + \tan\delta)\right)\right]^{3/2}}{2\left[B + (h_c - y)(\tan\beta + \tan\delta)\right]^{1/2}}. \tag{6.16}$$

Assuming $\beta = \delta$ we write a dependency between the lateral and vertical erosion as

$$\frac{\Delta B}{\Delta y} = Const$$

where ΔB and Δy represent the change in width (bottom or top) and in depth of the breach opening.

6.1.2 Second formulation of dam breaching development

In this formulation, a triangular breach channel is initially formed and progresses to the dam bottom (see Figure 6.1). The area of the triangular cross section of the breach channel (Figure 6.1), for $\beta \neq \delta$ is:

$$A_b = \frac{(H_d - y)^2(tan\beta + tan\delta)}{2}. \tag{6.17}$$

We differentiate A_b with respect to time t as follows

$$\frac{dA_b}{dt} = \frac{dA_b}{dy}\frac{dy}{dt} = -(tan\beta + tan\delta)(H_d - y)\frac{dy}{dt}. \tag{6.18}$$

The discharge through the breach channel is calculated as

$$Q_b = \left(\frac{g}{2}\right)^{1/2}h_c^{5/2}\frac{\tan\beta + \tan\delta}{2}, \tag{6.19}$$

where, the critical depth is

$$h_c = \frac{4}{5}(H - y). \tag{6.20}$$

Combining the Eqs. 6.6, 6.8 and 6.18, the breach bottom elevation can be estimated every time step with

$$\frac{dy}{dt} = -\frac{1}{2}\frac{k_{er}}{k^3}(\gamma g)^{3/2}\frac{h_c^2}{H_d - y}\left(\frac{\cos\beta + \cos\delta}{sin(\beta + \delta)}\right)^{3/2}. \tag{6.21}$$

Once the dam bottom is reached, the breaching develops in lateral direction into a trapezoidal shape. The breach area cross section for trapezoidal breach shape with $y < 0$ is equal to

$$A_b = (H_d - 2y)(\tan\beta + \tan\delta)\frac{H_d}{2}. \tag{6.22}$$

The wetted area is

$$A = (h_c - 2y)(\tan\beta + \tan\delta)\frac{h_c}{2}, \tag{6.23}$$

where h_c is the critical depth derived as

$$h_c = \frac{3y + 2H + \sqrt{9y^2 - 8Hy + 4H^2}}{5}. \tag{6.24}$$

Using the above expressions the discharge through the breach channel may be written as

$$Q_b = \sqrt{\frac{g}{2}}\frac{\tan\beta + \tan\delta}{2}[h_c(h_c - 2y)]^{3/2}(h_c - y)^{-1/2}. \tag{6.25}$$

The breach bottom elevation can be estimated at every time step with

$$\frac{dy}{dt} = -\frac{k_{er}}{k^3}(\gamma g)^{3/2}\frac{h_c^2}{2H_d}\frac{(h_c - 2y)}{(h_c - y)^{3/2}}\left[\frac{\cos\beta + \cos\delta}{\sin(\beta + \delta)}h_c - y\right]^{1/2}\frac{\cos\beta + \cos\delta}{sin(\beta + \delta)}. \tag{6.26}$$

Based on the formulations described above, a software package (see Figure 6.2) is developed using Borland Delphi. The program has user–friendly graphical interface, which allows the user to load the necessary information and to run the model.

6.2   Validation of the BREADA model

We validate the BREADA model against a historical dam failure event: the breaching of the Schaeffer Dam in the USA that failed due to overtopping in 1921. To extend even further the comparison with the real measurements (or known information), we also compare our results to the output of the commercially available model BREACH (Fread, 1988). We analyze the breach outflow time series generated by both models and compare the peak outflow values with the results obtained from empirical equations (introduced in Section 4.3) and data mining techniques (introduced in Section 5.3).

The Schaeffer Dam was a 30.5m high non–cohesive earthfill dam constructed across Beaver Creek in USA. The dam failed due to heavy rains described in Follansbee and Jones (1922) as ‘cloud-bursts’, which led to water levels in the reservoir higher than the dams top elevation. Some of the dam and reservoir characteristics are presented in Table 6.1.

According to Follansbee and Jones (1922), the dam was washed away in 30 minutes.

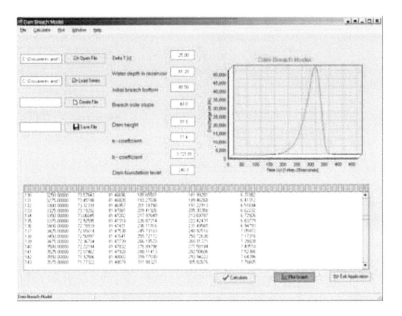

Figure 6.2: Main window of the BREADA program.

Table 6.1: Schaeffer Dam characteristics.

Dam Characteristics	
Dam height [m]	30.5
Embankment width at crest [m]	4.6
Embankment length at crest [m]	335
Upstream slope [H:V]	1:3
Downstream slope [H:V]	1:2
Grain size D_{50} [mm]	2.5
Uniform factor d90/d30	12.4
Porosity	0.5
Unit weight [KN/m^3]	18.5
Friction angle	25 - 40
Hydraulic Characteristics	
Reservoir [m^3]	3.92 million
Initial water depth in the reservoir [m]	30.5
Inflow [m^3/s]	270
Initial breach depth [m]	0.3

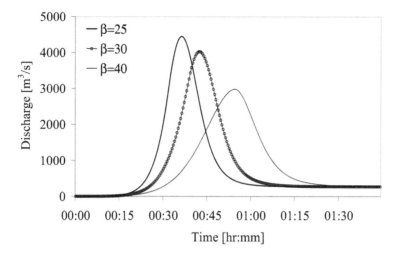

Figure 6.3: Sensitivity analysis for breach side angle with the BREADA model.

At the end of failure the breach opening was observed to be of a trapezoidal shape with a top width of about 210m (some 70% of dam crest length) and side slopes of approximately 1:2.25 [H:V]. A peak outflow of 4,500m^3/s was recorded.

The available record of parameters for the Schaeffer Dam is incomplete and, therefore, a sensitivity analysis is undertaken to identify which of the (unknown) input data influences the outflow discharge. The inflow into the reservoir was estimated by Follansbee and Jones (1922) to be 270m^3/s and belongs to the maximum discharge estimated for Beaver Creek, due to the storm that occurred more than 24 hours before the failure of the dam. It includes the rainfall and tributary inflow. In our experiment a reservoir stage–area curve is created based on the available information for some representative reservoir levels. Model parameters, namely the initial breach depth, breach side slope and erodibility coefficient are part of the sensitivity analysis for the BREADA model and the internal friction angle is part of the sensitivity analysis for the BREACH model. In this section, we briefly discuss the influence of side slope and internal friction angle on breach outflow hydrograph.

Figure 6.3 presents the resulting outflow hydrograph simulated for sensitivity analysis of the breach channel side slope ($\beta = \delta$) by the BREADA model. The angle that the breach sides create with the vertical is varied from 25° to 40°. It can be observed that the variation in the side slope influences the peak outflow magnitude and its time of occurrence. This is expected since the breach channel area depends on the breach channel side slope. Smaller is the angle that the breach side creates with the vertical, smaller is the initial breach channel cross section area. This leads to smaller initial breach outflow meaning an increase of water depth in the reservoir.

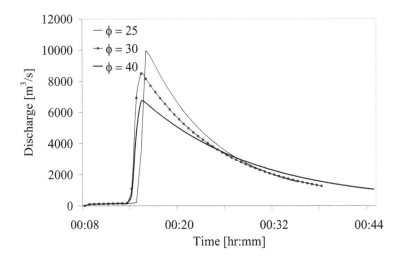

Figure 6.4: Sensitivity analysis for internal friction angle with the BREACH model.

This process continues for few minutes. Then the water depth in the reservoir that is increased as a result of lower outflow values, initiates higher erosion rates, thus higher discharge rates flowing through the breach channel. In Figure 6.3 we can observe that the angle of 25° is producing a more realistic breach outflow hydrograph, with peak values similar to the observed one and a flood duration of about 30 minutes. This result is consistent with the observed final side slope of the breach channel of 1:2.25.

Figure 6.4 presents the sensitivity analysis of the BREACH model for variable internal friction angle. The angle of 40° generates a peak outflow about 40% higher than the recorded value, but lower than the peak outflow obtained for angles of 25° and 30°. For the internal friction angle equal to 60°, the obtained side slope angle is 26° coinciding with the observed one and the corresponding peak outflow is 4,000m³/s. This happens due to the same reason as for the BREADA model. The internal friction angle value affects the enlargement of the breach width. During all the simulations, the duration of the hydrograph rising limb in the BREACH model is very short, fast reaching the maximum peak outflow.

In order to evaluate the applicability of data mining techniques to dam breach modelling, the models created with IBk and MLP algorithms in Section 5.3.3 are used to predict the peak outflow for the Schaeffer Dam and the resulting peak outflow values are analyzed. The results of peak outflow estimation for the Schaeffer Dam are presented in Figure 6.5. The peak outflow estimated by the IBk model is 4,000m³/s viz. about 10% smaller than the observed value, while a 34% lower peak outflow is predicted by the MLP model.

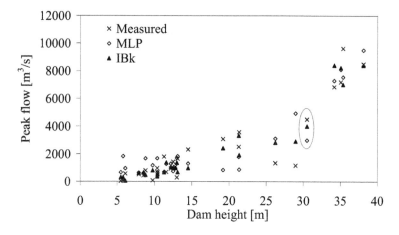

Figure 6.5: Peak outflow estimation based on dam height.

Table 6.2: Peak outflow estimation using empirical equations for Schaeffer Dam.

Reference		Peak outflow (m^3/s)
Regression Equations		
Singh and Snorrason (1982)	- Eq. 4.5	2230
MacDonald and Langridge-Monopolis (1984)	- Eq. 4.4	2584
Costa (1985)	- Eq. 4.11	2422
Froehlich (1987)	- Eq. 4.12	3843
Envelope Equations		
Hagen (1982)	- Eq. 4.2	9700
MacDonald and Langridge-Monopolis (1984)	- Eq. 4.3	8460
Costa (1985)	- Eq. 4.7	6991
	- Eq. 4.8	9435

Using the empirical relations (see Table 6.2), the peak outflow varies from $2{,}230m^3/s$ to $3{,}843m^3/s$ excluding the upper limits from the envelope equations. Froehlich's equation estimating peak outflow based on depth and volume of water in the reservoir at failure time, performs better than the other equations, where the estimated peak outflow is 15% lower than the recorded peak outflow.

6.3   Failure analysis of an earthfill dam

Failure of large dams is most likely to result in the release of enormous quantities of water, endangering people's lives and property in downstream areas. The conse-

Figure 6.6: View of the Bovilla Dam facing upstream.

quences are particularly high in terms of human casualties when no warning system is available. A dam break analysis is required for every constructed large dam for the purpose of identifying the impact of failure in the downstream areas and the necessary measures for mitigating the consequences.

In this section, we undertake a dam break analysis for the hypothetical failure of Bovilla Dam (Figure 6.6) that was constructed on the Terkuze river, 15km northeast of Tirana City, Albania. It is an earthfill homogeneous dam, 81m high, and 135m long at the crest. The dam's filling material is composed of gravel and sand, taken from alluvial deposits of the Terkuze River upstream of the dam. The normal capacity of the reservoir (Figure 6.6) is 80 million m^3 at the pool elevation of 318m from which 50 million m^3 are used for water supply and the rest for irrigation. Bovilla Reservoir can accommodate the average annual volume that the Terkuze River brings, that is evaluated to be nearly 105 million m^3 water. The purpose for constructing the Bovilla Dam is to provide enough water to the continuously expanding population of Tirana city and to satisfy the demands for irrigation purposes.

Though dams are meant to function for a long period of time, the probability of their failure is never zero. The Bovilla Dam is classified into the high hazard category due to populated areas located just 10 km from the dam. The dam construction lead to permanent displacement of more than 400 families from their homes and traditional livelihood in 1993. However, a year later, people started to settle near

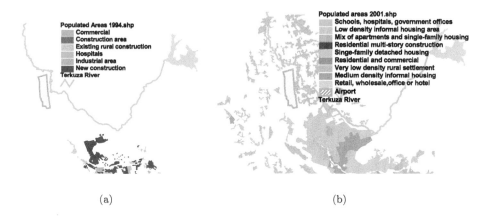

Figure 6.7: Land use data for the area downstream the dam showing the populated areas for the years (a) 1994 and (b) 2001.

the river banks downstream the dam. Figure 6.7 shows the land use data for two different years: 1994, a year after the work for construction of the dam had been started and 2001, five years after the work was completed. A potential failure of the Bovilla Dam may lead to highly devastating consequences in the areas downstream. Therefore, a dam break analysis needs to be conducted to identify the potential flood propagation scenarios and consequences in case of a failure event. Modelling of the flood wave propagation induced by a potential dam break event can help the implementation of emergency plans, risk assessment, and future development of the areas.

The pertinent properties of the dam and the reservoir are presented in Table 6.3. The impenetrability of the upstream slope of the dam is achieved by mean of a geo-textile material. There is a bottom outlet, used for water release for flood control and irrigation purposes. The total discharge capacity of the bottom outlet is $150\text{m}^3/\text{s}$ and is designed to handle 1,000 year return period flood event with peak discharge in the river of $700\text{m}^3/\text{s}$. This flood brings an increase of the water level in the reservoir to maximum of 320m. Two gates are installed at the outlet tower: the principal gate and the emergency gate. Next to the outlet tower an intake tower releases water on a regular basis for water supply.

The hydrologic data consist of the flood discharge time series for return period events of 10, 100 and 1,000 years (Albinfrastrukture, 1996). The relationship between reservoir surface area and water elevation (Figure 6.9) is obtained from digital elevation data (Figure 6.8) using Arc–GIS. An area 32km downstream of the dam is considered in this analysis. The terrain elevation varies from 240m above sea level at the

Table 6.3: Bovilla Dam Pertinent Data.

Dam Characteristics	
Dam top elevation, [m a.s.l]	321
Height above streambed [m]	81
Length at the crest [m]	135
Width of top crest [m]	8
Upstream slope	1.6
Downstream slope with 4 berms	1.6
Reservoir Characteristics	
Normal pool elevation [m a.s.l]	318
Maximal pool elevation [m a.s.l]	319
Normal pool area [km^2]	4.1
Normal pool capacity [m^3]	75 million
Maximum pool capacity [m^3]	84.3 million
Catchment area [km^2]	98
Bottom Outlet	
Type	armoured concrete surface
Bottom elevation at the entrance (m a.s.l)	275
Principal gate dimension [m]	3.5×3.7
Emergency gate dimension [m]	3.5×4.0
Length of outlet gallery [m]	156
Maximum capacity [m^3/s]	150
Freeboard	
Normal pool [m]	3

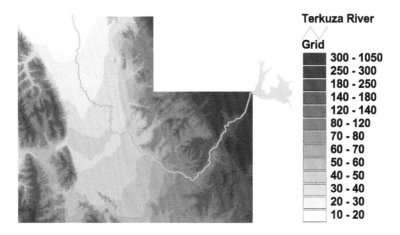

Figure 6.8: Digital elevation grid for the area downstream the dam.

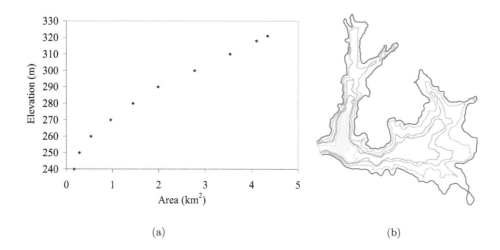

(a)   (b)

Figure 6.9: (a) Surface area-water elevation relationship for Bovilla reservoir and (b) Surface boundary for different elevations starting from inner boundary at 240m, 250m, 270m, 280m, 300m, 321m elevation, respectively

dam site to 15m above sea level at the downstream boundary. The floodplain at the first kilometers from the dam is quite narrow and steep, but expands rapidly afterwards.

6.3.1   Modelling of breach development

The first step in this dam break analysis is the prediction of the breach outflow hydrograph that will constitute the upstream boundary condition of the flood routing model. The main goal is the modelling of the flood wave propagation in case of the dam's failure, regardless of the expected probability of occurrence of such event. Here, we describe in detail the overtopping failure of the dam that might represent the worst flooding scenario (high magnitude of flood conditions) and discuss the piping failure.

Different scenarios can lead to dam overtopping, namely the inflow into the reservoir is higher than the spillway's capacity, malfunctioning of the spillway, etc. Currently, there is no methodology that is able to determine the extent of overtopping (flow depth and duration of flow over the crest) that the earth dam can withstand. Dam failure depends on the resistance of the embankment soil to water flow, type of dam covering material, slope of the embankment, which in turn influences the flow velocities, etc. In this study, we conduct a sensitivity analysis for the inflows into the reservoir assuming the 1,000 year and the 100 year return period flood events to inflow into the reservoir when the dam erosion is initiated. We ignore the effect of

other small stream and rainfall inflows into the reservoir and the hydrologic characteristics of the basin.

Other assumptions used in modelling are the following:

- Final breach bottom level corresponds to the stream bed elevation. Failure
  of a high dam, might not always develop down to the bottom as the width
  of the dam is usually much wider at that location. However, we make this
  assumption for the sake of simplicity.

- Water depth in the reservoir at the initial time is assumed at the dam top
  elevation or 0.2m above.

We start with a sensitivity analysis of the two different breach development formulations: (a) trapezoidal breach shape throughout simulation (see Section 6.1.2), and
(b) triangular breach shape till dam foundation is reached and trapezoidal breach
shape afterwards (see Section 6.1.1). In Figure 6.10 we can observe that the hydrographs do not differ significantly from each other in terms of peak outflow magnitude, but differ in timing. The trapezoidal breach development formulation exhibits
slower breaching of the dam. While this might seem counterintuitive, the result can
be expected due to the fact that the ratio between the vertical and lateral erosion is
limited to a constant factor (in this case $\Delta B/\Delta y = 2$), signifying slow enlargement
of the breach width. For the following simulations we use the second formulation of
breach development unless specified otherwise.

Figure 6.11 presents the sensitivity analysis for two different inflows into the reservoir. It can be observed that the magnitude of inflow into the reservoir does not
influence notably the flood hydrograph. The peak of the flood hydrograph is reached
about 6 minutes faster and is only 3% higher for the 1,000 year return period flood
event compared to the 100 year return period event. That is expected as the inflow
peak discharge for the 1,000 year return period flood is just $700\text{m}^3/\text{s}$ compared to
$430\text{m}^3/\text{s}$ for the flood with return period of 100 years, with both values being significantly lower than the breach outflow. For further analysis we assume the inflow
in the reservoir at the failure time to be 1.05 times higher than the peak discharge
of 1,000 year return period flood used for the design of the spillway, thus considering dam (and the spillway) to be able to accommodate lower inflow floods into the
reservoir.

Two different stage–volume relationships are created for the reservoir: one based on
the GIS data and the other based on the data provided by the designers of the dam
(Albinfrastrukture, 1996). The latter does not include the reservoir dead storage
information, but only the active storage capacity. The peak outflow obtained by the
model for different reservoir stage–volume relationships differs as much as 15% (see
Figure 6.12). This emphasizes the uncertainty related with the actual volume of

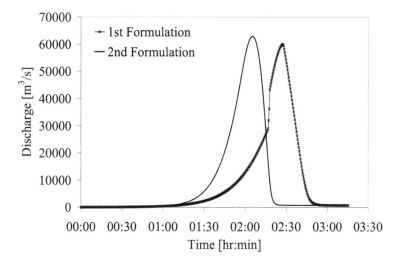

Figure 6.10: Breach outflow hydrograph for first and second breach development formulations.

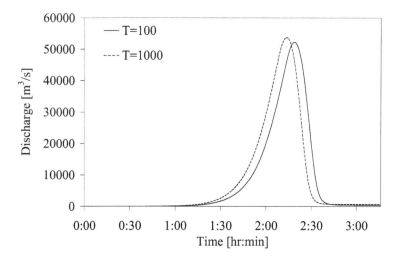

Figure 6.11: Breach outflow hydrograph for 1,000 and 100 year return period flood event.

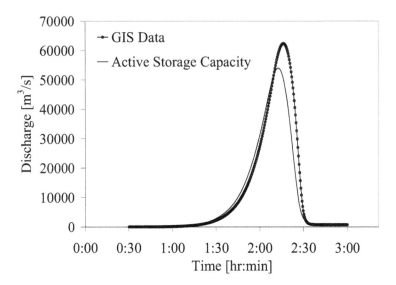

Figure 6.12: Breach outflow hydrograph for two reservoir stage–volume relationship.

water in the reservoir at the failure time, which is affected as well from the sediment
mass accumulated in the basin throughout the years, though not considered in this
analysis.

Two user–defined parameters of the BREADA model, initial breach channel depth
D and the angle that the breach channel sides create with the vertical $\beta = \delta$, are
subject to a sensitivity analysis. Different values for initial breach channel depth
(D=1m, 0.5m, and 0.3m) are considered in order to understand the influence of
the initial conditions on the breach development. As shown in Figure 6.13, D does
not affect the breach outflow magnitude, but has an impact on the peak timing.
Higher initial depth in the breach opening results in faster breach development. In
contrast, β has an influence on the breach outflow hydrograph (Figure 6.14). For
different values of β the magnitude of the peak discharge changes. An increase of
β leads to a decrease of the peak discharge and vice versa similarly as observed in
Section 6.2. Compared to the angle of $35°$, the magnitude of the breach outflow for
angles of $40°$ and $45°$, decreases 11.8% and 15.8% respectively.

Furthermore, we compare the dam breach modelling results obtained from the
BREADA and BREACH model. In Figure 6.15 it can be observed that though the
peak outflows differ in range of only 10%, the timing, which is a very important factor for proper warning of the population downstream from the dam, is significantly
different. In the BREACH model the peak outflow occurs after about 39 minutes
from the beginning of initial breach, while the peak outflow in the BREADA model

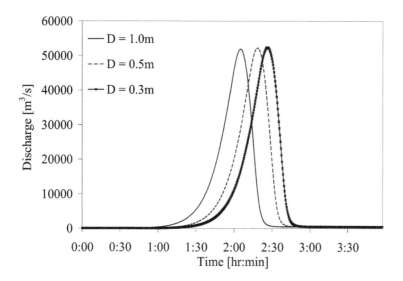

Figure 6.13: Sensitivity analysis for different initial breach depth D (β=40°; $Q_{1\%}$).

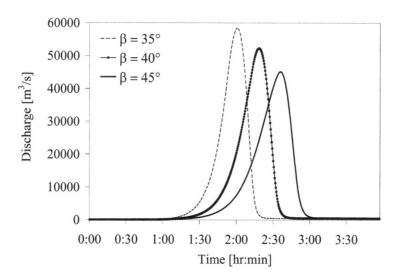

Figure 6.14: Sensitivity analysis for different breach angle β (D=0.5m; $Q_{1\%}$).

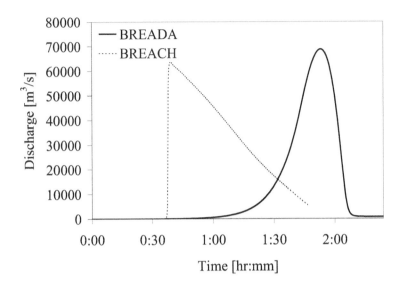

Figure 6.15: Breach outflow hydrograph for the BREACH and BREADA model.

occurs about 70 minutes later. A sharper rising limb is produced by the BREACH model compared to a more gradual development of the breach in the BREADA model.

The breach cross section could be a possible reason for this discrepancy - the BREACH model develops a rectangular breach shape from the beginning till the moment when the sides of the breach channel collapse, creating a trapezoidal shape. This moment in time depends on the dam material internal friction angle that in turn controls the side slope. Instead, the BREADA model develops a triangular shape till the bottom of the dam is reached and then develops into a trapezoidal breach shape. This leads to a slower development of the breach in the latter model. However, the duration of the rising limb produced by the BREACH model is only 6 minutes, and that might not be realistic. The results obtained from both breach models represent two different scenarios, one being less catastrophic to the downstream area due to slower development despite 10% higher peak outflow, and the other predicting a disastrous situation with faster development of breaching. In both models we do not take into account the cover layer at the upstream slope of the dam, which might influence the results. Peak outflows obtained by the BREACH and BREADA models are 63,570m^3/s and 68,740m^3/s respectively.

To conclude, we predict the peak outflow using the empirical equations. From the results in Table 6.4, it can clearly be seen that only Hagen's formula produces almost the same peak outflow as predicted using the dam breach models. The upper curve

Table 6.4: Peak outflow estimation based on dam height and reservoir storage.

Reference		Peak outflow (m^3/s)
Hagen (1982)	- Eq. 4.2	67,230
MacDonald and Langridge-Monopolis (1984)	- Eq. 4.3	43,630
	- Eq. 4.4	13,030
Costa (1985)	- Eq. 4.9	38,910
	- Eq. 4.11	13,950
	- Eq. 4.7	39,560
	- Eq. 4.8	58,950
Froehlich (1987)	- Eq. 4.12	31,980
Broich (1998)	- Eq. 4.13	14,260
	- Eq. 4.14	21,040

of the envelope equation proposed by Costa is 17% and 9% lower than the peak outflow produced by the BREADA and BREACH models respectively. These results were expected since Hagen's formula is the only formula derived from historical data of earthfill dams failing due to overtopping, with a range of characteristics that covers the characteristics of the Bovilla Dam. The dam considered in this study, is higher than all of the 22 dams that Froehlich used for deducing his equation. Moreover, other empirical formulas were extracted from data sets where no distinction was made between dam material or failure type.

6.3.2   Modelling of flood propagation

The breach outflow hydrograph generated from the BREADA and BREACH models is routed in the areas downstream the dam using Sobek 1D2D, an integrated one– and two–dimensional numerical simulation package developed by WL | Delft Hydraulics (see Section 3.4). A 1D model is created for the 32km reach downstream of the dam with an average bed slope of 0.01. The 2D modelling area is approximately 19km×17km with grid cell size of 30m×30m. Since the purpose of this analysis is the identification of flooded areas in case of a hypothetical failure of the Bovilla Dam, rather than the identification of the flood duration period, the simulation is finalized when the flood wave reaches the downstream boundary.

We ignore the reservoir sediment or the dam material, which are flushed away together with the water stored in the reservoir, but instead consider propagation of *clear* water in the valley. The flooded areas are treated as *rigid*, meaning neither erosion nor deposition processes are modelled despite the awareness that these processes will occur during a real dam failure flood wave propagation. The presence of the debris material in the flood water, which will affect the flow pattern significantly,

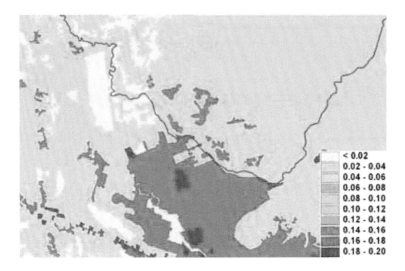

Figure 6.16: Roughness grid.

is taken into account through higher values of roughness coefficients.

The roughness of the floodplain area is subject to sensitivity analysis. The extent of the dam break flood event is unprecedented in the natural history of the valley, thus the calibration of the roughness coefficients is not possible. Sobek 1D2D has an option to apply constant or variable roughness coefficients for the channel reach and the overland area. A constant Manning's roughness coefficient n equal to 0.07 is assumed for the river channel. Constant and variable roughness coefficients are used for the overland area. The roughness coefficients for different land use categories (Figure 6.16) are deduced from the Tables of Manning's n presented in Chow (1959); van der Sande et al. (2003); Huang (2005).

The flood wave propagation scenario for the breach outflow hydrograph generated by the BREADA model is presented in Figure 6.17. The extent of flooding for two time intervals is given for the purpose of comparing with the flood patterns when the breach outflow hydrograph of the BREACH model is used as upstream boundary condition in the flood modelling analysis. Simulation results show that the flood wave reaches the location 26 km downstream of the dam (downstream of the river reach presented in Figures 6.17 and Figure 6.18) in approximately 5, 4, and 3 hours time span for constant Manning $n = 0.15$, $n = 0.1$, and for roughness based on land use data. It reaches the populated areas located 9 km downstream the dam in 1hr and 40 minutes, enough time to warn population for immediate evacuation. At 13 km away from the dam, flood peaks are dampened out as the wave reaches the wide flat areas; still, the water depth at the villages, including the international airport

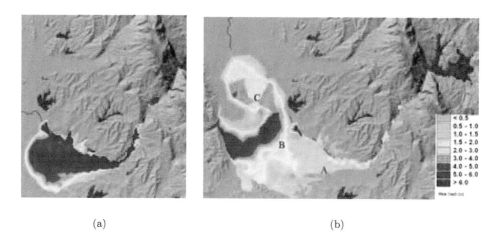

(a)   (b)

Figure 6.17: Routing of the BREADA model outflow hydrograph using Sobek 1D2D with Manning coefficient in the floodplain area equal to 0.15 (a) 2hr 38min after the breach has been initiated and (b) 4hr 38min after the breach has been initiated.

exceeds 2m. The flood speed reaches 7m/s at the populated areas when taking into account the land use based roughness.

The flood wave propagation scenario for the breach outflow hydrograph generated by the BREACH model is presented in Figure 6.18. Owing to the fast breach development in the BREACH model, the flood wave reaches 26 km downstream of the dam in 4, 3, and 2 hours time span for constant Manning $n = 0.15$, $n = 0.1$, and for land use based roughness. In this scenario the time interval from the initial breaching to the time the flood wave approaches the first residential areas is only 40 minutes. Higher water depths are observed in some locations in comparison to the BREDA model scenario (see Figure 6.19).

The different roughness maps provide different flood hazard scenarios. The results of a sensitivity analysis for constant and land use based Manning roughness coefficients on the floodplain are presented in Figure 6.19. The water depths are shown for various locations in the valley downstream the dam (see Figure 6.17). As expected, the increase in floodplain roughness coefficient decreases the flood propagation speed, thereby delaying the flooding of a particular area. An increase in water depth is observed at the high elevation locations (e.g. location A which correspond to a residential area) while there is no notable change in water depth at the low elevation areas (location C that corresponds to the airport area). The water depth in A obtained using the BREACH model hydrograph as upstream boundary condition is lower than the one obtained by using the BREADA model hydrograph, associated

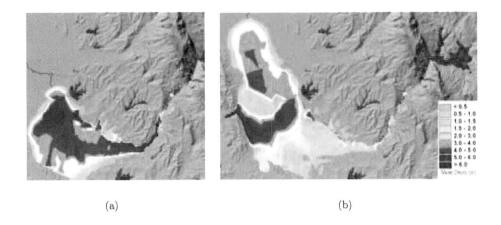

(a)   (b)

Figure 6.18: Routing of the BREACH model outflow hydrograph using Sobek 1D2D when Manning coefficient in the floodplain area equals to 0.15: (a) 2hr 18min after the breach has been initiated and (b) 4hr 18min after the breach has been initiated.

this with the higher peak outflow values for the latter. These differences are dampened at the location C further away from the dam where the floodplain area is wide.

According to the flood maps the water depths in residential areas are in the order of 5m. The failure of Bovilla Dam would jeopardize the lives of thousands of people living downstream and extensively affect the environment and the economy of the area. Based on preliminary population data for the regions inundated by flood water, the population at risk consists of at least 10,000 people. In case of a dam failure event, the population faces significant risk. Immediate construction of a flood warning system is necessary to avoid human casualties, in particular for residents living as close as 9km away from the dam.

The inundation maps provide an estimate regarding the direct consequences, but the indirect (long term) consequences as a result of dam failure will be enormous as well. The development of the areas should be planned by taking into consideration the above mentioned flood hazard maps ensuring better protection of the people facing the risk in a potential dam failure event, as well as decreasing economic and environmental consequences as a result of the failure.

6.4   Uncertainty in dam break analysis

When physically based methods are used for forecasting the flood wave propagation of a dam failure event, no data are usually available for its validation. Instead the focus is on the identification of sources of uncertainty and the range to which

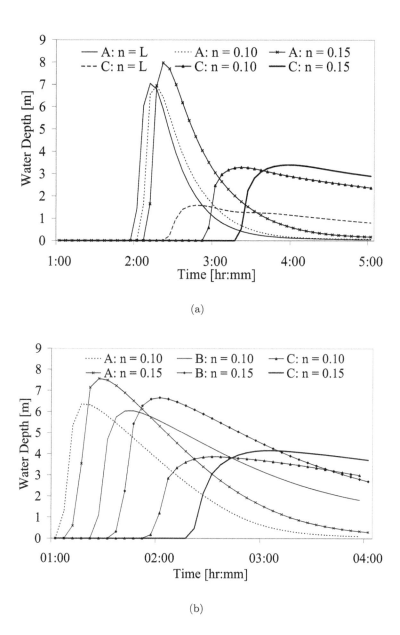

(a)

(b)

Figure 6.19: Sensitivity analysis of the roughness coefficient at locations 11km, 15km, and 19km downstream the dam (namely A, B, and C respectively) for (a) BREADA outflow hydrograph (L stands for land use data) and (b) BREACH outflow hydrograph.

they affect the model results. Identifying the uncertainty associated with the prediction of floods imposed by dam failure is necessary for effective flood management and emergency planning. Three sources of uncertainty are identified in Section 3.4 namely, input uncertainty, model uncertainty, and completeness uncertainty. The latter, often referred as *ignorance* uncertainty, is associated with all omissions that occur due to lack of knowledge.

Input uncertainty

Neither the inflow into the reservoir nor the water depth in the reservoir at the failure time are known for the analysis of a hypothetical dam failure. Both values are assumed depending on the event that will be simulated e.g. inflow is higher than the spillway design capacity or the dam design flood for the overtopping failure, and the water depth in reservoir corresponds to the maximum depth. These characteristics are a source of uncertainty. However, the sensitivity analysis shows that inflow hydrograph influence is not significant for the large Bovilla Dam. An increase in the inflow with 60% (from $Q_{1\%}$ to $Q_{0.1\%}$) for Bovilla reservoir resulted in an increase of the peak outflow with only 1%.

Although the reservoir area and storage volume are considered to be deterministic quantities, the calculated reservoir area and storage volume at a given elevation may vary due to different measurement and computational techniques used for their estimation. Moreover the reservoir area and storage volume might be different from what is assumed, depending on the sedimentation in the reservoir during the years before the failure occurs. The reservoir characteristics, namely the approximated shape, volume, and area, all affect the final results.

Dam material characteristics at the failure time are part of the sensitivity analysis and are very important for the correct estimation of the breach development and characteristics. Roughness coefficient and Digital Elevation Model (DEM) data for the floodplain area influence the scale of flooding in terms of magnitude and time and are part of sensitivity analysis.

Model uncertainty

Model uncertainty is associated with different formulations implemented for modelling the processes involved during the breaching of the structures, mathematical and numerical description of the processes, etc. Different models result in different output for the common problem due to various approximations made in the description of the processes. The model related parameters influence significantly the scale of flooding. We observe that two formulations of the BREADA model, using different breach shapes for breach development, lead to different breach outflow hydrographs with differences in timing, which is a very important factor in flood analysis. The same is observed when comparing the BREACH and the BREADA models.

Completeness uncertainty

The completeness uncertainty represents unknown contributions and hence is a qualitative analysis rather than quantitative. Processes that are currently considered irrelevant for the dam failure flood analysis, might prove to be important if investigated in greater detail. There are many processes that we omit in modelling due to current lack of knowledge or awareness (e.g. wind can have impact on the failure of the dam but its influence is usually not considered in any analysis for high dams except for dikes). For the overtopping of an earthfill dam, the wind direction might influence the failure process. The wind blowing toward the dam during high flood might contribute to higher depth of water overtopping the dam.

6.5 Discussion and conclusions

This chapter starts by describing the model developed for predicting the breach characteristics of an earthfill dam failing due to overtopping. A sensitivity analysis is undertaken for assessing the influence of the model parameters on the computed breach outflow hydrograph. It is observed that the initial breach channel depth, that is a user defined parameter, has an influence on peak timing. The breach side slope affects both the peak timing and its magnitude.

Two different methods are used for predicting breach outflow hydrograph in case of hypothetical failure of an earthfill dam. We compare the resulting breach outflow hydrographs obtained from the BREADA (Zagonjolli et al., 2005) and the BREACH model (Fread, 1988). Despite only 10% difference in the peak outflow values obtained from the two breach models, significant differences are observed in timing and shape of the hydrograph.

Furthermore, we compare the resulting peak outflows of the BREACH and BREADA models with the range of peak outflows obtained using the empirical formulations developed during the last decades and data mining techniques (see e.g. Zagonjolli and Mynett (2005b,a)). Though empirical formulations can be used for a 'rough/fast' prediction of the peak outflow values, they are not applicable for dam break flood forecasting where knowledge about breach development in time is important. Hagen's empirical formula produces almost the same peak outflow as predicted using physically based numerical models. Other empirical formulas do not show good agreement, that can be explained from the fact that the Bovilla Dam characteristics are not within the range of dam characteristics in the datasets used for developing those formula.

The resulting breach outflow hydrographs, constituting the upstream boundary condition for the 1D2D hydrodynamic model, led to two different flood progression scenarios in the areas downstream from the dam. For numerical simulation of flood propagation downstream of the dam, the breach models are coupled with the Sobek 1D2D modelling package developed by WL | Delft Hydraulics (the Netherlands). The choice of erodibility factor for the BREADA model turns out to have a high influence on the breach outflow hydrograph timing and shape, but here we assume a scenario where the breach development is slower, simulating flood propagation for the fastest (BREACH) and slowest flood wave (BREADA). Roughness in the areas subject to flooding influences flood speed and characteristics. A constant and variable roughness based on land use data is used.

Based on preliminary population data for several regions inundated by flood water, the Population at Risk consists of at least 10,000 people. Immediate construction of a flood warning system is proposed, as a key source for avoiding loss of human lives, for residents living as close as 9km away from the dam; since in case of a dam failure event, this population faces significant risk.

Chapter 7

A Numerical-Constraint based Model

The more constraints one imposes, the more one frees one's self. And the arbitrariness of the constraint serves only to obtain precision of execution.

Igor Stravinsky

7.1 Introduction

During past decades different methodologies were proposed to prevent and manage floods. The traditional approach to prevent the impact of flooding through flood protection is more recently being replaced by a flood management approach (Plate, 2000; Brinkhuis–Jak et al., 2003; de Vriend, 2005; Samuels et al., 2005; Simonovic and Ahmad, 2005) recognizing that absolute flood prevention is unachievable and unsustainable, due to high costs and inherent uncertainties. In the Netherlands, the standard policy of raising dikes crest levels in order to maintain the required level of flood protection is being abandoned in favor of a new policy of creating 'Room for the River' that involves widening river cross sections by relocating dikes further away from the river, lowering floodplains, etc. (van Schijndel, 2005). Partitioning of the area at risk into compartments, utilizing the highway and railroad embankments as well as the natural terrain, could possibly lead to an increase in warning time and a reduction of the flooded area.

The risk associated with flooding is usually defined as the product of probability and consequences. This expression leads to a situation where the total risk of an event with low probability and high consequences is equal to the total risk of an event that has high probability of occurrence but low consequences. Generally, risk reduction measures aim to reduce the probability of flooding and might often be easier to implement, although minimizing the probability of a flood can come at the expense of increasing its destructive power (when it happens), thus increasing the consequences.

Simulation of the consequences of a predicted or assumed flood is typically elaborated in what–if scenarios. Every what–if scenario requires specification of the initial state and the configuration of the system. Considering different scenarios makes the number of required simulations grow increasingly large due to the possible combinations of choices for the initial state and the configuration of the system. Therefore, in practice, only a small fraction of all possible strategies can be explored.

In this chapter we present a numerical–constraint based approach for flood risk reduction and decision support as well as a case study illustrating its application to a system of polders at risk (Zagonjolli et al., 2006). The model developed is able to simulate different flood mitigation scenarios taking into account the social and economic value of the areas that could be prone to inundation. The impact of flooding is minimized by selecting the most feasible mitigation scenarios.

The hydrodynamic modelling packages require extensive and lengthy calculations, while the proposed 'lightweight' numerical–constraint based technique offers advantages of simultaneous evaluation of different flood mitigation scenarios, taking into account different economic and social aspects that a traditional numerical system is not able to use. In this chapter we demonstrate the possibility of integrating both approaches, viz. the usage of the developed method for providing the most feasible mitigation scenario and detailed hydrodynamic modelling that can be carried out at a later stage. This in turn allows for inclusion of the proposed approach, into any existing hydrodynamic modelling package.

7.2   Description of the numerical–constraint based approach

The key elements of the developed numerical–constraint based technique are the *graph algorithms* that are applied to simulate different possible flood propagation scenarios occurring in areas prone to inundation. Thus, a flood mitigation problem is converted into a problem of finding the most feasible path for water propagation, which can be addressed within the existing framework of graph algorithms (Corman et al., 2001).

Figure 7.1 shows the transformation of a flood prone area into a graph. The area surrounded by outer dikes (either *wet* or *dry* boundary dikes) consists of 12 polders separated by inner dikes (dikes that belong to more than one polder). Each of the polders, represented as a vertex v of the graph, has a maximum capacity to store

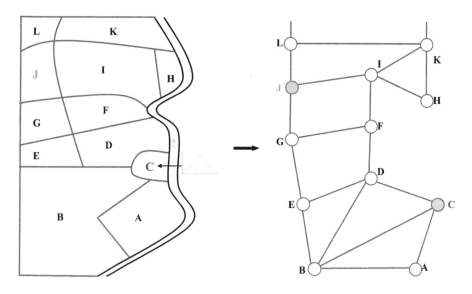

Figure 7.1: Transformation of a system of polders into the graph.

flood water. For each inner dike there is an edge connecting two vertices (polders). Therefore, all possible ways of flood propagation in the area prone to flooding are represented as *all possible paths from the initially flooded polder to the polder that needs to be protected*. For example, we can assume an initial breaching to occur at the river dike in polder **C** as a result of hydraulic conditions or as a result of the decision to deliberately initiate breaching at that particular location. We also assume that polder **J** is estimated (based on the given evaluation) to be the most important polder that requires maximum protection. Then the algorithm finds the optimal route for the flood water propagation avoiding the most valuable polder(s) or a particular polder that needs to be protected. Therefore, one of the main questions arising while addressing a flood mitigation problem, that is *find all possible paths for flood water before it arrives at the protected area*, is efficiently addressed by our numerical–constraint based model, utilizing graph algorithms with different selection criteria as discussed later in this chapter.

A notable advantage of this approach over traditional numerical models is the calculation of a single (or small number) of paths found by the graph algorithm to be most feasible flood propagation scenarios, resulting in minimal damage. Using a 'forward approach' involving traditional numerical simulation would require a 'brute' force approach of evaluating all (many) possible ways of the flood propagation, thus conducting extensive simulations and only later providing the user with the possibility to manually select the most appealing scenario. This might lead to unnecessary complications during the simulations, resulting in high computation cost and complexity of the model.

We distinguish two ways of modelling flood propagation by:

1. Ignoring economic, social and environmental values of the area, thus reflecting a 'naturally' occurring flood propagation scenario; or

2. Considering economic, social and environmental values of the area prone to flooding and proposing a preferred route for the flood water to follow.

Another important feature of the proposed method is that different constraints can be taken into account for social, economic and environmental values of the areas prone to flooding. The model is able to determine the most valuable polder(s) to be protected, based on user supplied information. While the value of economic assets can be determined with some degree of certainty, it is much more difficult to accurately estimate the value of non–economic assets such as environmental, cultural and social considerations. Objects having national significance can be taken as an important constraint for inundation of a particular area, and - together with the population at risk - will contribute to the overall score of the polder's value.

Let us denote the directed graph in Figure 7.2 as $G = (V, E)$ where V is the vertex set corresponding to the available polders and E is the edge set and any edge corresponding to the to link between adjacent polders. In case of directed graphs every edge $(u, v) \in E$ is represented through a set of ordered pairs of vertices. In our case, the direction of each of the edges is not decided beforehand, but according to the hydraulic conditions in each polder at the every time step, as explained later in the chapter. We distinguish two types of vertices in our graph: a *source* s and a *sink* t. If polder **C** is the initial flooded polder then this polder will be the *source* vertex in the graph network and polder **J**, the most valuable polder in the area which we aim to protect from flooding, will be the *sink* vertex. Other vertices in the graph that are neither *sources* nor *sinks* are called *intermediate* vertices. Water flows through the intermediate vertices and might be stored in them. We assume that every vertex lies on some path from the source to the sink vertex. The rate at which water enters a vertex must equal the rate at which it leaves the vertex and the amount remaining in the vertex, so mass conservation is preserved.

A non–negative real–valued function p is defined on the edges set and represents the capacity function of our network G. Its value on a particular edge is the capacity of the edge e that represents the maximum rate of flow from one polder in the other. The capacity function of the network is related to the initial flow conditions at polder **C**.

The flow in the network should satisfy the following constraints

$$0 \leq g(e) \leq p(e), \ \forall \ e \in E \tag{7.1}$$

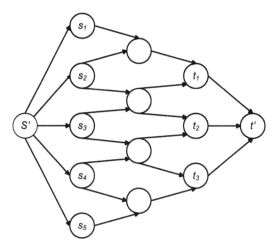

Figure 7.2: A network with multi source and multi sink vertices.

and

$$g^-(v) = g^+(v), \ \forall \ v \in V. \tag{7.2}$$

The value $g(e)$ expresses the rate at which the water flow travels along e. The upper bound in condition (7.1) is called the *capacity constraint*; it imposes the natural restriction that the flow along an edge cannot exceed the capacity of the edge or cannot be higher than the available water in the polder from where water is flowing out. Condition (7.2), called the *conservation condition*, requires that for any intermediate vertex (polder), the volume of water flowing into v is equal to the volume flowing out of v. The conservation condition implies that the flow entering Polder **C** is equal to the sum of the volumes of water distributed over the intermediary and sink vertex. Our aim is, given the network G with *source s* and *sink t*, to find the best way of accommodating a flood in the paths consisting of polder vertices that have a low economic value. As polder **J** is the polder to be protected from being flooded, we assume that the graph has many *sinks* that are the polders adjacent to the polder to be protected, and name polder **J** the *supersink t*.

Inundation of areas is based on mass conservation (continuity equation):

$$\frac{dV_p}{dt} = Q_{in} - Q_{out}, \tag{7.3}$$

where V is the polder volume, t is time and Q_{in} and Q_{out} are the flow rates in each direction into and out of the cell (see Figure 7.3), corresponding to flow through the breaching dike.

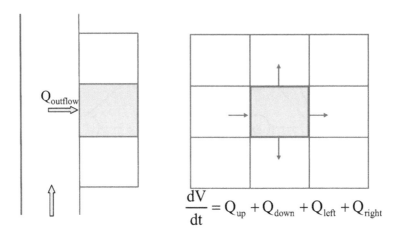

Figure 7.3: Mass conservation principle in an area compartmentalised in many polders. Each cell in the domain represents a polder.

There are different factors that might initiate the failure of a dike structure as described in Section 2.2. PC–Ring (Vrouwenvelder, 1999; Vrouwenvelder et al., 2001) is a tool developed and used in the Netherlands to calculate the total failure probability of a flood defence system consisting of dikes. A dike ring system is as weak as its weakest link (dike section). Thus, the failure probabilities are calculated for the weakest links in a system of dikes and not for every dike section. The total failure probability includes the failure probabilities for overtopping and overflowing, instability of inner or outer slope, uplifting/piping. The failure probabilities are usually calculated for dikes along a river, sea or other water body; it is common that the failure probability of dikes that are not exposed to water, i.e. tertiary dikes is not known.

Currently, in our model, we consider the overflowing failure mode as the most likely failure mode for dikes during a flood event. During overflowing, the water level exceeds (even in the absence of waves) the crest level of the dike and the water flowing over the dike and along the inner (landward) slope induces structural failure. The probability of water levels exceeding the dike crest level is expressed as $P(Z < 0)$, where Z is a reliability or state function that represents the difference $Strength - Load$. For $Z < 0$, the load is higher than the strength of the dike and as such it fails, and vice versa (for $Z > 0$ the dike does not fail). The state when $Z = 0$ is called the limit state. For overflowing of the dikes, Z is expressed as follows:

$$Z = h_d + \left(\frac{q_c^2}{0.36g} \right)^{1/3} - h_w, \tag{7.4}$$

where h_d is the dike height [m], q_c is the critical discharge [m^3/s] or the maximum discharge flowing over the dike for which failure of the dike is not initiated, and h_w is the occurring water level [m]. The second term in Eq. 7.4 expresses the critical depth on a broad crested weir.

The flow exchange between two polders in case of a dike failure from overflowing is estimated using the BREADA model (Zagonjolli et al., 2005; Zagonjolli and Mynett, 2006a,b) presented in Chapter 6 that is adapted for dike failure by increasing the erodibility coefficient and taking into account the submergence of the breach flow. Since the BREADA model is applicable for non–cohesive dams, that means that all the dikes in the domain are assumed to be non–cohesive dikes.

The mutual dependence between the safety of different compartments (polders or dike ring areas) comes from hydraulic and economic interactions as summarized in the following:

- Due to high discharge in the river, deliberate failure of one of the dike sections can be induced, which need not be the weakest dike section along the river. This deliberate failure leads to flooding of a less valuable area while it might lead to the attenuation of the flood wave in the more valuable areas downstream, preventing other potential failures.

- When two polders or dike ring areas share a common dike, its failure leads to flooding of both adjacent areas.

- Failure of one dike section might prevent failure of another dike section due to a decrease of the water level in the polder.

- From an economic point of view: inundation of one area can impose damage to another non–inundated area due to *economic links* (e.g. infrastructure, etc.).

The hydraulic interaction between compartments (dike ring or polder areas) influences the protection level of the area that might depend on another one. Furthermore, risk reduction in one particular area might be followed by an increased risk of flooding in the other one. Thus, a detailed assessment has to be made, including the interaction between different compartments. Within the framework of the proposed approach the interactions between compartments of a dike ring are taken into account. To minimize flooding impact on the complex hydraulic system of dike ring areas, different objectives could be taken into consideration during the simulation process:

1. Natural flooding: flooding of the compartments naturally occurring under specific hydraulic conditions

2. Longest Path: intentional flooding of compartments for accommodating the water, that is finding the longest path for the flood water propagation before it reaches the protected polder.

3. Maximum Storing Volume Capacity: selection of polders to be flooded based on their maximum volume capacity.

4. Minimal Total Damage: selection of polders to accommodate the flood water based on their socio–economic, cultural and environmental value. Less valuable polders to be flooded first.

5. Combination of the above objectives (2-4).

Below we describe in more detail how these objectives can be satisfied within the proposed methodology.

Natural flooding

We define the flooding of the area that occurs without deliberate intervention as occurring naturally. In a system of polders it is usually preferable to take into account the probability of several dike failure modes as well as the associated uncertainties. Knowing the failure probability of all dike sections is beneficial when identifying the potential breach location. Moreover, when deliberate breaching of a dike is considered, it provides the possibility to choose the dike section with highest failure probability. Our modelling tool is built to take into consideration this information, if provided by the user. However, in our example application we consider only dike elevation and water depth in the polder as the primary source for breach initiation. Once the water level in the polder raises higher than the top elevation of any of the surrounding dikes, overflow breaching of that dike is assumed to be initiated leading to the flooding of the polder. Another important element for dike breach initiation is the duration of the overflowing water, which should be large enough to initiate erosion. We do not specify the overflow duration time in our model, but ensure that the duration is long enough as the filling up of the first polder coincides with the rising limb of the inflow hydrograph, which implies that further increase of water level in the polder is expected.

Longest path flooding

Probably one straightforward way to prevent flood water reaching the most valuable polder, is the requirement to accommodate as much water as possible along the flood path. That is, to select the longest path from the source polder to the protected one. Our numerical–constraint based model employs for that purpose a graph algorithm that generates all possible paths between them and selects the longest possible one. The algorithm returns a list of vertices (polders), including the start and the end vertices (polder **C** and **J**) comprising the path. The same vertex does not occur more than once in the returned path (no cycles are allowed). In some cases the algorithm can return more than one unique longest path, if several are available. In these situations, our model randomly selects a single path or provides the user the possibility to select a preferred one.

Then the numerical model for natural flooding uses the suggested path to simulate controlled flood propagation. This means that previously not breached dikes will be breached for the purpose of accommodating the excess water and controlling flood propagation to protect the most valuable polder. In case the flood could be retained within the polders located on its path before reaching most valuable polder, then **J** will be dry, otherwise it will be the last polder to be flooded, thus having more time for evacuation.

Maximum storing volume capacity flooding
Another possibility to defend the most valuable polder as well as to cause minimal damage to the rest, is to accommodate the flood water in the polders with maximum storage capacity. Each of the polders in the domain has a maximum storage capacity which is assumed to be equal to the polder area multiplied by the minimum height of its surrounding dikes. The algorithm tries to find the best solution for accommodating the flood water in the polders that have higher storage capacities. For this purpose we again utilize a graph algorithm that finds the path, such that its storage volume capacity is maximal. Then the numerical–constraint based model uses this path to accommodate flood water and to protect the most valuable polder from flooding, in the same way as discussed previously.

It can be observed that in some cases both the longest path and the maximum storage volume objectives could lead to the same solution. Obviously, in case of a unique longest path, it possesses the largest storage capacity than any other one. However, the situation is completely different in case a restriction on the number of polders that should be used to accommodate flooding is imposed.

Minimal total damage flooding
Finally, the most economically feasible flood propagation scenario could be imposed by enforcing a minimal total damage constraint. The key idea behind this approach is to identify and deliberately flood areas that have the least economic value. Thus, controlled flooding is aimed to produce the least total damage although deliberately flooded areas might have low flooding probability. The graph algorithm finds a path along the least valuable areas in the domain based on the user defined input data in terms of socio–economic, cultural and environmental value. Below we present an application of our method using synthetic data to demonstrate its main features and usability.

7.3 Application

The domain area of the case study considered in this section is schematized in Figure 7.4 together with the information about dike height and top elevation as well as the polder's maximum storage capacities and economic values. The value is expressed in

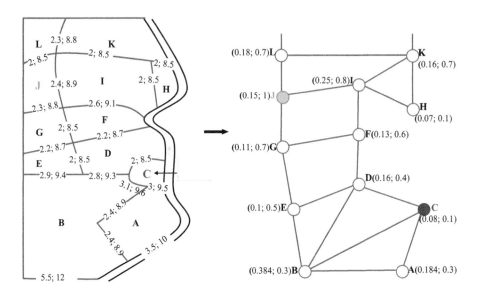

Figure 7.4: Transformation of the case study area into the graph. Dike height and the dike top elevation [m] as well as polder area [km^2] and its relevant value are presented.

relative terms varying in a scale from 0 to 1. However, the overall score is calculated based on a particular economic value of each polder and its social and environmental aspects are considered as contributing elements, although not expressed in monetary terms. These values are not depending on the water level or velocities in the polder, but are used as indicators for the polders that are more valuable and require most protection. Including a depth–damage function as one of the constraints in our model is foreseen as future work and can be incorporated into existing framework.

TAW (2002) provides guidelines for the critical discharge on overtopped sea dikes of clay or sand material with either good or bad grass cover. However, to our knowledge, no information exists concerning the critical discharges of overflowed inner dikes. In the case study, $q_c = 0.16$m^3/s is assumed , which leads to a water depth of 0.2m above the dike crest to initiate dike failure. Due to the fact that the application is used for demonstration purposes of the developed method, we may argue that our assumption can be considered reasonable for the purpose of dike breach initiation. The initial dike to breach is the river dike of Polder **C**. The breach outflow for that dike is assumed rather than estimated using a breach model. The hydrograph has maximum peak outflow of 400 m^3/s. As can be observed in Figure 7.4 the border dikes have higher heights to satisfy the property of rigid boundaries that never suffer failure. The same condition is applied for the other river dikes as well.

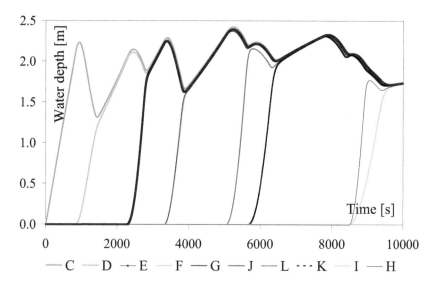

Figure 7.5: The natural flooding of the polders - water depth in polders as a function of time.

In Figure 7.5 the rate of change of water depth in every polder for a natural propagation of the flood in the domain is shown. It can be observed that Polder **J** is flooded. The breaching of the dikes is initiated when the water depth is higher than $h_d + 02m$. For simplicity, our domain is considered flat with bottom elevation of the dikes to be the same.

Using the Longest Path objective, for the area presented in Figure 7.4, the longest path suggested by the algorithm is [C, A, B, D, E, G, F, I, H, K, L, J] (see Figure 7.6 and Figure 7.7). Note, that the first and the last polders are considered as most crucial for the model, however the intermediate polders might not necessarily be flooded based on their ranking order in the path. We observe that in this case the polder we aim to protect is dry at the end of the simulations. The flood water is directed towards the other polders and the result is the expected flood propagation scenario at the end of the simulation.

The path obtained when selecting the polders with the largest maximum volume capacity is [C, A, B, D, E, G, F, I, H, K, L, J] which corresponds to the one obtained for the longest path. This could be expected since the longest path and the maximum volume objective appeared to be the same objectives in terms of storage capacity.

A very different flood path propagation (compared to the longest and maximum storage paths) is obtained for the minimum total damage scenario, thus selecting

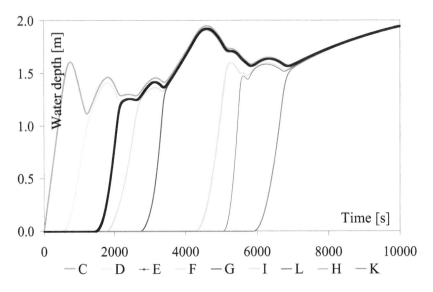

Figure 7.6: The longest path and the maximum polder volume flooding scenarios.

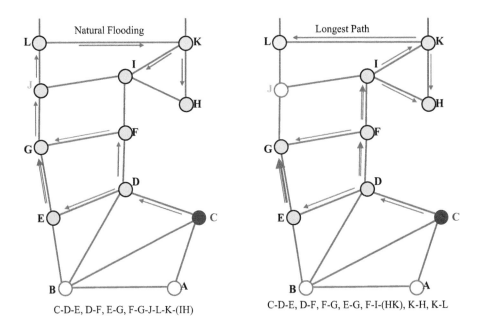

Figure 7.7: The natural and longest path flooding of the polders in graph representation.

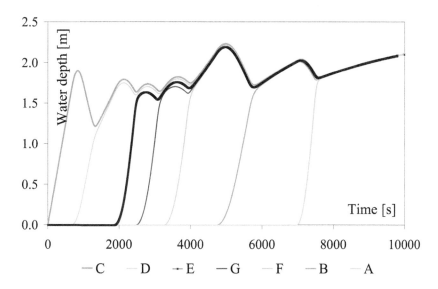

Figure 7.8: The minimum damage flooding scenario.

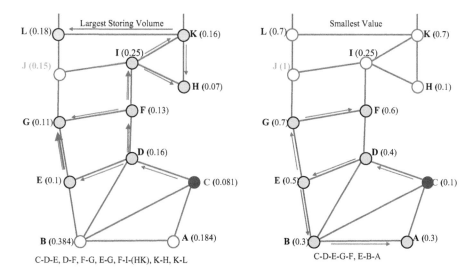

Figure 7.9: The flooding of the polders based on the largest storage volume and the minimum value objective.

the less valuable polders in our domain to be flooded. The polders are found to be [C, A, B, E, D, F, G, J] and the flooding pattern is shown in Figure 7.8 and Figure 7.9. The Polder **J** is dry at the end of this simulation.

7.4   Conclusions

The main idea behind the proposed approach of using graph theory to develop constraints for numerical simulations is to *manage* the flooding process within a system of polders or dike ring areas divided into a number of compartments. For this purpose we use a numerical–constraint based model that takes into account not only physical processes involved in dike breaching and flow propagation but also social, economic and environmental values of the areas prone to inundation. We simulate different scenarios of flooding based on the hydraulic processes and the constraints introduced to minimize the flood damage in each protected area.

By temporarily storing the water in compartments, the total risk might be minimized. The flooding probability is increased for some compartments, while the total flood risk is decreased. The filling process of the compartments depends on the failure process of the dike and on the condition whether a deliberate breaching of the dike is involved or 'natural' failure of hydraulic structure occurs. In both cases the same breaching model is used, but in the former case dikes are deliberately lowered. There is a possibility to include faster breach development of the dikes that might not have a slow breaching process depending on the methods used to remove (open) a dike section. Furthermore, the model can be enhanced by including the presence of additional hydraulics structures.

The key difference of our method in comparison to existing ones, is that instead of assigning a particular probability to a scenario, we propose an alternative approach of evaluating and mitigating consequences based on a numerical–constraint based approach. In particular, we evaluate different ways for flood mitigation during a flood event. In the simulation we take into account compartments created by the available dikes, roads and natural high grounds in the areas prone to flooding. Through constraints, we optimize the strategy for choosing the most feasible flood propagation scenario that minimizes socio–economic consequences. This is done by utilizing algorithms based on graph theory that efficiently transform dike ring areas into a graph and find different optimal paths applicable for each objective function. The main strategy consists of navigating floodwater 'away' from the most populated or economically valuable regions.

Based on the considerations presented above, we may suggest that flood risk management could draw significant benefit from a constraint based methodology by using a set of measures that 'keeps the system dry'. Including targets and goals inside the support system, allows better prediction of different scenarios since in this case the model will provide optimal solutions. However, the case might be that some extreme floods in the low-lying areas, caused by prolonged heavy rainfall might not be prevented as a result of measures taken or the method used.

Chapter 8

Conclusions and Recommendations

. . . the river . . .
Keeping his seasons and rages, destroyer, reminder
Of what men choose to forget. Unhonoured, unpropitiated
By worshippers of the machine, but waiting, watching and
waiting.

T.S. Eliot

8.1   Conclusions

In this thesis, various approaches are proposed and applied for simulating dam and dike failure events, flood water routing in downstream areas, and flood risk reduction, providing a unified framework for addressing flood related events. Numerical, data mining, graph theory and constraint based methods are applied to the problem of breach modelling and flood water mitigation. The findings related to each of the proposed approaches are listed below.

8.1.1   Traditional and novel approaches for predicting dam breach characteristics

For every dam or dike there is a limit to which the structure is built to withstand the forces applied. Unexpected or unforeseen events might trigger failure, when the strength of the structure is weaker than the acting load. In this thesis a range of modelling techniques is explored to deal effectively with the failure event of hydraulic structures such as earthfill dams and dikes.

We focus on dams and dikes composed of earth material since they constitute the largest percentage of total number of dams and dikes built around the world and the largest number of failures occurs in these structures as well. For structures with cohesive material, two processes are most likely to develop: erosion and headcutting. The latter was found to be predominant during field and laboratory experiments (Morris, 2005). The cohesive bond between the particles of the cohesive soil presents higher resistance to the flow, while for the non–cohesive soil the physical properties of the particles (size, shape, density, porosity, and fall velocity) contribute to the resistance to the flow. Because the particles are not bound together, the erosion process is believed to be the predominant breaching mechanism. However, any instability in part of the structure might also result in mass failure. In this thesis, a dam breach model is developed for simulating overtopping failure of non–cohesive dams.

BREADA model

The BREADA model developed in this thesis assumes two different breach shape evolutions in time and space with two important time intervals. In the first formulation, the initial trapezoidal breach shape develops in vertical and lateral direction, until the ground is reached and then progresses in lateral direction only. In the second formulation, a triangular breach shape is initially assumed and continues to expand in vertical direction until the bottom level of the structure is reached. During the second phase, the trapezoidal breach progresses in lateral direction until either the water flow force can not cause any further erosion or the breach has reached its maximum allowed dimensions that are limited by the dam geometry. Flow through the breach channel is calculated as a weir flow and the erosion of the dam material is estimated with a bed load transport empirical equation. The model developed is validated against the historical failure event of Schaeffer Dam in USA. Good agreement is observed in terms of peak outflow prediction for a range of model parameters.

Data mining models

We apply data mining techniques for deriving dam breach characteristics and peak outflow from available data. These methods generally rely on large amount of data, but in the applications presented in this thesis, the information about historical dam failure events is sparse. Still, performance of data mining techniques is found to be slightly better than the current empirical equations, suggesting that enriching the databases with more data about real dam failure events is one possible way to improve the models that can further be used to evaluate the capabilities of the physically based methods by providing the range of peak outflow and breach characteristics or to provide some of the input information needed for the simulation of the semi–physically based models.

Uncertainty in dam breach modelling

The uncertainty associated with each of the processes involved in forecasting flood risk occurring due to breaching of structures is identified and quantified in relative terms. We distinguish between model uncertainty and input uncertainty.

Four different methods to calculate dam breach peak outflow are evaluated against the documented failure event of the Schaeffer Dam. We compare the breach outflow hydrograph and the peak outflow obtained from our model with the BREACH model as well as with the range of peak outflows obtained from empirical equations and data mining techniques. The results show a wide range of predicted peak outflows. When using empirical equations, the peak outflow discharge varies in a range of ±50%. Though the best performing empirical equation for the Schaefer dam can be identified, this does not imply its direct applicability to other failure events. When the dam breach models BREACH and BREADA are used the predicted peak outflow varies in a narrower range depending on the assumptions and selected parameters. Good results are obtained from the IBk data mining model that predicts only a 10% lower peak outflow than the recorded one.

Many dam material characteristics are needed for modelling, but their (precise) values are not always available. The largest uncertainty is observed in the hydrograph shape, particularly in the timing of the breach development. While the BREACH model exhibits a very sharp rising limb meaning fast development of the breaching, a smooth rising limb is observed for the BREADA model, the former leading to worst downstream flood conditions while the slower wave produed by the latter will leave more warning time to the population downstream.

8.1.2   Identification of dam failure flood extent, uncertainty, and risk reduction

We simulate the hypothetical failure of the Bovilla Dam, an earthfill dam of 81m high and observe the impact of two outflow hydrographs obtained from the BREADA and BREACH models, in the downstream area. In terms of flood duration, the fast rising breach outflow hydrograph leads not only to faster flood propagation, but also to higher water depths in the flat areas further downstream the dam. The slow developing flood wave leaves room for longer warning period for the population and the lower water depth values in the flat areas, but higher water depths are observed in the first kilometers from the dam. The models, one leading to a less catastrophic scenario than the other, are examples of the uncertainty that current breach modelling techniques possess with respect to the most important element - timing. Peak outflows of the models do not differ much, and there is a good agreement with the empirical equation predictions.

Other sources of uncertainty also affect the modelling of dam break flood propagation, especially when failure of large dam is involved. The roughness of the areas subject to flooding is not known beforehand. The resulting flood propagation scenarios depend on the roughness coefficient used (either constant or land use based one). Moreover, we ignore the reservoir sediment or the dam material, which are flushed away together with the water stored in the reservoir, but instead consider propagation of *clear* water in the valley. The presence of the material attached to the water, which will affect the flow pattern significantly is taken into account through higher values of roughness coefficients.

Once the potential risk is identified, the task is to find the ways of reducing this risk and take measures for warning the population downstream in case of a structural failure event. Our modelling results can be used as flood hazard maps and can assist communities in planning future developments in areas that are prone to flooding.

8.1.3 A numerical–constraint based approach for flood mitigation

The flood risk associated with the failure of a high dam with a large volume of water stored in the reservoir and the flood risk associated with the failure of a river dike during a high discharge event is highly dependent on the vicinity of the structure to the populated areas and the measures taken for dealing with potential flooding. Here, we introduce a numerical–constraint based model for flood mitigation in low–lying areas that are subject to flooding during a flood event. Instead of assigning a particular probability to a scenario, we propose an alternative approach of evaluating and mitigating consequences based on a numerical–constraint based model. During the simulation we take into account compartments created by the presence of dikes, roads and natural high grounds in the areas prone to flooding. Through constraints, we optimize the strategy for choosing the most feasible flood propagation scenario that minimizes socio–economic consequences. This is done by utilizing graph based algorithms that efficiently transform compartmentalized areas into a graph and find different optimal paths applicable for each objective function. The strategy consists of navigating floodwater 'away' from most populated or economically valuable regions.

We believe that flood risk management could benefit from a constraint based approach by using a set of measures that keep the system dry. Including objectives in a hydrodynamic modelling package, allows better prediction of different scenarios since in this case the model will provide optimal solutions. However, it might still be the case that some extreme floods or floods caused due to prolonged heavy rainfall over the domain area might not be prevented as a result of measures taken or the method used.

8.2 Recommendations

The following recommendations for further research are suggested.

- During the past three decades different dam breach modelling methods have been proposed for predicting the outflow hydrograph and peak outflow value. However, none of the methods can guarantee to obtain accurate breach characteristics. The complexity of breach development usually motivates researchers to make many approximations to obtain simpler mathematical description of the processes involved, thus increasing the uncertainty related to the structural failure development. This is the case when modelling erosion and headcut erosion processes. Many laboratory and field experiments have been performed to provide in depth understanding of these processes. However, the complete physical description is far from perfect. A possible direction for future investigation could be the analysis of erosion processes in a river stretch during a high flow and evaluating its application to a dam breaching event.

- For high dams the monitoring and recording of failure events has not been satisfactory so far, making the calibration of breaching models difficult and leading to uncertainty in the prediction results. The drawback of the experimental tests is that the weakest spot on the dam structure is predefined, influencing further development of the breaching. The optimal solution is real time dam monitoring and aerial pictures of breach development, but these proposed measures might be difficult to implement, since breaching caused by overtopping usually occurs during bad weather conditions, therefore limiting the monitoring possibilities. However, due to technological advances both in hardware and software sensor technology, one might expect (real time) monitoring of dikes and dams during breach development to become possible and feasible.

- Scale models are currently the main source of validation data for dam/dike breach modelling. Although hydraulic conditions, such as water velocity, etc. are effectively replicated in laboratories, the exact scaling factors for the soil properties and sediment transport are usually not available. Therefore, an interesting direction for future research is to investigate scaling of soil related properties. Furthermore, simulation and visualization of breaching processes in 3D might lead to better descriptions of the phenomena.

- Improving the accuracy of existing physically based models by using the information extracted from a database of recorded dam failure events is another direction of this research. Currently, the available database consists of about 100 dam failure events and suffers from inconsistencies between the different sources. Increased efforts in documenting and sharing the information related to historical dam dike failure events is necessary. This will lead not only to better validation of the physically based models, but also to more accurate data mining models for discovering patterns and relationships among dam failure events.

- During structural breaching, the dam material is flushed away together with the water, changing the morphology in the areas downstream. The modelling of flood propagation is usually accomplished by assuming *clear* water and rigid terrain throughout the flood water propagation domain. However, including the interaction of debris flow and non–rigid terrain boundaries may significantly affect the flooding pattern. This problem has recently received significant attention and further research in this area is necessary.

- The numerical–constraint based approach proposed in this thesis can be extended by introducing different objective functions that are applicable for specific scenarios. Furthermore, its capabilities can easily be boosted by using an advanced numerical engine of any well known hydrodynamic package. On the other hand, it is also possible to include the constraint–based part of the method into existing hydrodynamic packages, which are often part of the decision making process for proposing measures needed to protect important areas in terms of their socio–economic and cultural value.

Bibliography

Abbott, M. B.: 1979, *Computational hydraulics: Elements of the theory of free surface flows*, Pitman Publishing Ltd., London.

Albinfrastrukture: 1996, Diga e Bovilles (in albanian and italian), *Technical report*, Ministry of Construction, Albania.

Alcamo, J. and Bartnicki, J.: 1987, A framework for error analysis of a long-range transport model with emphasis on parameter uncertainty, *Atmospheric Environment* **21**(10), 2121–2131.

Alonso, C. V., Bennett, S. J. and Stein, O. R.: 2002, Predicting head cut erosion and migration in concentrated flows typical of upland areas, *Water Resources Research* **38**(12), 39.1–39.15.

Archer, D.: 2006, *Global warming: Understanding the forecast*, Blackwell Publishing Ltd.

Asselman, N. E. M. and Middelkoop, H.: 1995, Floodplain sedimentation: Quantities, patterns and processes, *Earth Surface Processes and Landforms* **20**(6), 481–499.

Babovic, V., Keijzer, M., Aguilera, D. R. and Harrington, J.: 2001, An evolutionary approach to knowledge induction: Genetic programming in hydraulic engineering, ASCE, Proc. of the World Water and Environmental Resources Congress, Orlando, Florida, USA, pp. 64–64.

Birkes, D. and Dodge, Y.: 1993, *Alternative methods of regression*, John Wiley, New York.

Bogárdi, J.: 1974, *Sediment transport in alluvial streams*, Akadémiai Kiadó, Budapest, Hungary.

Bowles, D. S.: 2001, Evaluation and use of risk estimates in dam safety decision making, Proc. of the United Engineering Foundation Conference on Risk-Based Decision-Making, Santa Barbara, California, USA, pp. 17–32.

Brinkhuis–Jak, M., Holterman, S. R., Kok, M. and Jonkman, S. N.: 2003, Cost benefit analysis and flood damage mitigation in the Netherlands, Paper for the ESREL Conference, Maastricht, the Netherlands, pp. 16–18.

Broich, K.: 1998, Mathematical modelling of dam break erosion caused by overtopping, Proc. of the 2nd CADAM Meeting, Universität der Bunderswehr, München, Germany. Available at www.hrwallingford.co.uk/projects/CADAM/.

Broich, K.: 2003, Sediment transport in breach formation process, Proc. of the 3rd IMPACT Workshop, Université Catholique de Louvain, Louvain–la–Neuve, Belgium. Available at www.impact-project.net.

Brown, C.: 1950, Sediment transportation, *Engineering Hydraulics* pp. 769–857.

Brown, R. J. and Rogers, D. C.: 1977, A simulation of the hydraulic events during and following the Teton Dam failure, Proc. of Dam–Break Flood Routing Model Workshop, U.S. Water Resource Council, Bethesda, Maryland, pp. 131–163.

Brown, R. J. and Rogers, D. C.: 1981, Users manual for program BRDAM, U.S. Bureau of Reclamation, Denver, Colorado.

Chanson, H.: 2005, Applications of de Saint–Venant equations and method of characteristics to the dam break wave problem, University of Queensland, Australia.

Chaudhry, M. H.: 1993, *Open-channel flow*, Prentice Hall, New Jersey.

Chow, V. T.: 1959, *Open channel hydraulic*, McGraw-Hill, New York.

Coleman, S., Andrews, D. and Webby, M.: 2002, Overtopping breaching of noncohesive homogeneous embankments, *Journal of Hydraulic Engineering* **128**(9), 829–838.

Commissie Rivierdijken: 1977, Rapport Commissie Rivierdijken (in Dutch). The Hague, the Netherlands.

Corman, T. H., Leiserson, C. E. and Rivest, R. L.: 2001, *Introduction to algorithms*, MIT Press, Cambridge, USA.

Costa, J. E.: 1985, Floods from dam failures, Open-file report, U.S. Department of the Interior Geological Survey, Denver, USA.

Costa, J. E. and Schuster, R. L.: 1988, The formation and failure of natural dams, *Geological Society of American Bulletin* **100**(7), 1054–1068.

Cristofano, E. A.: 1965, *Method of computing erosion rate for failure of earthfill dams*, U.S. Department of the Interior, Bureau of Reclamation, Denver, Colorado.

Cunge, J. A., Holly, F. M. and Verwey, A.: 1980, *Practical aspects of computational river hydraulics*, Pitman Publishing Ltd., London.

CUR/TAW: 1990, *Probabilistic design of flood defences*, Report 141, Technical Advisory Committee on Water Defences, Center for Civil Engineering Research and Codes, Gouda, the Netherlands.

de Saint-Venant, A. J. C.: 1871, Théorie du mouvement non-permanent des eaux, avec application aux crues des rivières et á lintroduction des marées dans leur lit, *Comptes Rendus des Séances de l'Académie des Sciences* **73**, 147–154.

de Vriend, H.: 2005, State of the art in flood management research, Proc. of the 3rd International Symposium on Flood Defence, Nijmegen, the Netherlands.

Deltacommissie: 1960, Rapport Deltacommissie (in Dutch), Staatsuitgeverij.

Dhondia, J. F. and Stelling, G. S.: 2002, Application of one dimensional–two dimensional integrated hydraulic model for flood simulation and damage assessment, *Proceedings of the 5th International Conference in Hydroinformatics* pp. 265–276.

Dodge, M. M.: 1997, *Hans Brinker or the silver skates*, The World Wide SchoolTM, Seattle, Washington, USA.

Dodge, R. A.: 1988, Overtopping flow on low embankment dams–Summary report of model tests, REC–ERC–88-3, U.S. Bureau of Reclamation, Denver, USA.

Dressler, R. F.: 1952, Hydraulic resistance effect upon dam-break functions, *Journal of Research, National Bureau of Standards, USA* **49**(3), 217–225.

Dressler, R. F.: 1954, Comparison of theories and experiments for the hydraulic dam-break wave, Vol. 3, Proc. of the International Association of Scientific Hydrology, Rome, Italy, pp. 319–328.

Dressler, R. F.: 1958, Unsteady Non-Linear Waves in Sloping Channels, *Proceedings of the Royal Society of London. Series A, Mathematical and Physical Sciences* **247**(1249), 186–198.

Du Boys, M. P.: 1879, Études du régime du Rhône et de laction exercée par les eaux sur un lit à fond de graviers indéfiniment affouillable, *Annales des Ponts et Chaussées* **5**(18), 141–195.

Edgeworth, F. Y.: 1887, On observations relating to several quantities, *Hermathena* **6**(13), 279–285.

Einstein, H. A.: 1942, Formulas for the transportation of bed load, *ASCE Transactions* **107**, 561–573.

Exner, F. M.: 1925, Ber die wechselwirkung zwischen wasser und geschiebe inflssen, *Sitzungber. Acad. Wissenscaften Wien Math. Naturwiss, Abt. 2A, 134* pp. 165–180.

Flannery, T.: 2006, *The Weather Makers: how man is changing the climate and what it means for life on earth*, Atlantic Monthly Press.

FLOODsite: 2005, Language of Risk-Project Definitions, Consortium Report T32-04-01. Available at www.floodsite.net.

Follansbee, R. and Jones, E. E.: 1922, The Arkansas River flood of June 3-5, 1921, *Water-Supply Paper 487*.

Franca, M. J. and Almeida, A. B.: 2004, A computational model of rockfill dam breaching caused by overtopping (RoDaB), *Journal of Hydraulic Research* **42**(2), 197–206.

Fread, D. L.: 1988, BREACH: an erosion model for earthen dam failures, Hydrologic Research Laboratory, National Weather Service, NOAA.

Froehlich, D. C.: 1987, Embankment-dam breach parameters, Proc. of the ASCE National Conference on Hydraulic Engineering, Williamsburg, Virginia, USA, pp. 570–575.

Funtowicz, S. O. and Ravetz, J. R.: 1990, *Uncertainty and quality in science for policy*, Springer.

Galy-Lacaux, C., Delmas, R., Kouadio, G., Richard, S. and Gosse, P.: 1999, Long-term greenhouse gas emissions from hydroelectric reservoirs in tropical forest regions, *Global Biogeochemical Cycles* **13**(2), 503–517.

Gauss, C. F.: 1809, *Theoria motus corporum coelestium in sectionibus conicus solem ambientium*, Vol. Translation reprinted as *Theory of motions of the heavenly bodies moving about the sun in conic sections*. Dover, New York, 1963, Perthes et Besser.

Gerritsen, H.: 2005, What happened in 1953? The Big Flood in the Netherlands in retrospect, *Philosophical Transactions: Mathematical, Physical and Engineering Sciences* **363**(1831), 1271–1291.

Hagen, V. K.: 1982, Re-evaluation of design floods and dam safety, Vol. 1, Proc. of 14th Congress of International Commission on Large Dams, Rio de Janeiro, Brasil, pp. 475–491.

Hanson, G. J., Robinson, K. M. and Cook, K. R.: 2001, Prediction of headcut migration using a deterministic approach., *Transactions of the ASAE* **44**(3), 525–531.

Hanson, G. J., Temple, D. M. and Cook, K. R.: 1999, Dam overtopping resistance and breach processes research, Proc. of the Annual Conference Association of State Dam Safety Officials, St. Louis, Missouri, USA.

Harris, G. W. and Wagner, D. A.: 1967, Outflow from breached earth dams, University of Utah, Salt Lake City, Utah, USA.

Hartford, D. N. D. and Baecher, G. B.: 2004, *Risk and uncertainty in dam safety*, Thomas Telford.

Hervouet, J. M.: 2000, A high resolution 2-D dam-break model using parallelization, *Hydrological Processes* **14**(13), 2211–2230.

Hervouet, J. M.: 2007, *Hydrodynamics of free surface flows: Modelling with the finite element method*, John Wiley & Sons.

Hoffman, F. O. and Hammonds, J. S.: 1994, Propagation of uncertainty in risk assessments: The need to distinguish between uncertainty due to lack of knowledge and uncertainty due to variability, *Risk Analysis* **14**(5), 707–712.

Hooijer, A., Klijn, F., Pedroli, G. B. M. and van Os, A. G.: 2004, Towards sustainable flood risk management in the Rhine and Meuse river basins: synopsis of the findings of IRMA-SPONGE, *River Research and Applications* **20**(3), 343–357.

Huang, Y.: 2005, *Appropriate modeling for integrated flood risk assessment*, PhD thesis, University of Twente, the Netherlands.

IMPACT: 2004, Investigation of Extreme Flood Processes and Uncertainties, EC Contract EVG1-CT-2001-00037, www.impact-project.net.

International Commission on Large Dams: 2005, *Risk assessment in dam safety management: a reconnaissance of benefits, methods and current applications*, CIGB/ICOLD. Bulletin 130.

IPCC: 2001, *Climate Change 2001: The scientific basis*, Third Assessment Report of the IPCC, Cambridge University Press.

Janssen, P. H. M., Heuberger, P. S. C. and Sanders, R.: 1994, UNCSAM: A tool for automating sensitivity and uncertainty analysis, *Environmental Software* **9**(1), 1–11.

Johnson, F. A. and Illes, P.: 1976, A classification of dam failures, *International Water Power and Dam Construction* **28**(12), 43–45.

Klijn, F., van Buuren, M. and van Rooij, S. A. M.: 2004, Flood-risk management strategies for an uncertain future: Living with Rhine River floods in the Netherlands, *AMBIO: A Journal of the Human Environment* **33**(3), 141–147.

Knight, D., Cao, S., Liao, H., Samuels, P., Wright, N., Liu, X. and Tominaga, A.: 2006, Floods are we prepared?, *Journal of Disaster Research* **1**(2), 325–333.

Krzysztofowicz, R.: 2001, The case for probabilistic forecasting in hydrology, *Journal of Hydrology* **249**(1), 2–9.

Kwadijk, J. and Rotmans, J.: 1995, The impact of climate change on the river Rhine: A scenario study, *Climatic Change* **30**(4), 397–425.

Lafitte, R.: 2001, *Hydropower*, The 20th edition of the Survey of Energy Resources, World Energy Council.

Leal, J., Ferreira, R. M. L. and Cardoso, A. H.: 2002, Dam-break waves on movable bed, Vol. 2, River Flow 2002, Proc. of the International Conference on Fluvial Hydraulics, Louvain-la-Neuve, Belgium, Swets & Zeitlinger, Lisse, the Netherlands, pp. 981–990.

Lou, W. C.: 1981, *Mathematical modeling of earth dam breaches*, PhD thesis, Colorado State University, USA.

Loukola, E. and Huokuna, M.: 1998, A numerical erosion model for embankment dams failure and its use for risk assessment, Proc. of the 2nd CADAM Meeting, Universität der Bundeswehr, München, Germany.

Macchione, F. and Rino, A.: 1989, Dimensionless analytical solutions for dam-breach erosion, *Journal of Hydraulic Research* **27**(3), 447–452.

MacDonald, T. C. and Langridge-Monopolis, J.: 1984, Breaching charateristics of dam failures, *Journal of Hydraulic Engineering* **110**(5), 567–586.

McCarthy, J., Minsky, M. L., Rochester, N. and Shannon, C. E.: 1956, A proposal for the Dartmouth summer research project on Artificial Intelligence, Retrieved from www-formal.stanford.edu/jmc/history/dartmouth.

McCulloch, W. S. and Pitts, W.: 1943, A logical calculus of the ideas imminent in nervous activity, *Bulletin of Mathematical Biophysics* **5**, 115–133.

McCully, P.: 1996, *Silenced rivers: The ecology and politics of large dams*, Zed Books, London & New Jersey.

Merabtene, T., Yoshitani, J. and Kuribayashi, D.: 2004, Managing flood and water-related risks: A challenge for the future, Proc. of the AOGS Joint 1st Annual Meeting & 2nd APHW Conference, Singapore.

Meyer-Peter, E. and Müller, R.: 1948, Formulas for bed-load transport, Proc. of the 2nd Meeting of the International Association for Hydraulic Structures Research, pp. 39–64.

Millennium Ecosystem Assessment: 2005, *Ecosystems and human well-being: Synthesis*, Island Press.

Minns, A. W. and Hall, M. J.: 1996, Artificial neural networks as rainfall–runoff models, *Hydrological Sciences Journal/Journal des Sciences Hydrologiques* **41**(3), 399–417.

Mitchell, T. M.: 1997, *Machine learning*, McGraw-Hill Higher Education.

Montuori, C.: 1965, Introduction d'un débit constant dans un canal vide, Proc. of the 11th IAHR Congress, Leningrad, Russia, pp. 1–7.

Morris, M. W.: 2005, IMPACT: Final technical report, Proc. of the 1st IMPACT Workshop, HR Wallingford, UK.

Mynett, A. E.: 2002, Hydroinformatics in aquatic resources management, Proc. of the NATO Advanced Research Workshop on "New Paradigms in River and Estuary Management", Moscow, Idaho, USA.

Mynett, A. E.: 2004a, Artificial Intelligence techniques in environmental hydroinformatics, Proc. of the IAHR-Asian and Pacific Division Conference, Hong Kong.

Mynett, A. E.: 2004b, Hydroinformatics tools for ecohydraulics modelling, Proc. of the 6th International Conference on Hydroinformatics, Singapore.

Mynett, A. E.: 2005, *Hydroinformatics applications in hydroscience and engineering*, Chapter 2, Encyclopedia of Hydrological Sciences, Willey Publishers.

Mynett, A. E. and de Vriend, H. J.: 2005, Next generation information and communication technology (ICT) for integrated water management of the Yellow River Basin, Proc. of the 2nd Yellow River Forum, Zhengzou, China.

Nishat, A.: 2006, Experiences from Bangladesh on disaster management with focus on floods, Meeting Report on Regional Journalist Workshop on Water Issues in Asia, Bangkok, Thailand.

NPDP: 2007, Dam incidents database, National Performance of Dams Program, Stanford University, California, USA. Available at npdp.stanford.edu.

Nsom, B.: 2002, Horizontal viscous dam–break flow: Experiments and theory, *Journal of Hydraulic Engineering* **128**(5), 543–546.

Olías, M., Cerón, J., Fernández, I., Moral, F. and Rodríguez-Ramírez, A.: 2005, State of contamination of the waters in the Guadiamar Valley five years after the Aznalcóllar spill, *Water, Air, & Soil Pollution* **166**(1), 103–119.

Plate, E. J.: 2000, Flood management as part of sustainable development, Vol. 1, Proc. of the International Symposium on River Flood Defence, Kassel, Germany.

Ponce, V. M. and Tsivoglou, A. J.: 1981, Modeling gradual dam breaches, *Journal of Hydraulic Division* **107**(7), 829–838.

Powledge, G. R., Ralston, D. C., Miller, P., Chen, Y. H., Clopper, P. E. and Temple, D. M.: 1989, Mechanics of overflow erosion on embankments. Part II. Hydraulic and design considerations, *Journal of Hydraulic Engineering* **115**(8), 1056–1075.

Price, R. K.: 2000, Hydroinfomatics for river flood management, Vol. 71, NATO Science Series 2 Environmental Security, Kluwer Academic Publishers, pp. 237–250.

Pugh, C. A. and Harris, D. W.: 1982, Prediction of landslide generated water waves, Vol. Q. 54, Proc. of the 14th International Congress on Large Dams (ICOLD), Rio de Janerio, Brazil, pp. 1056–1075.

Qing, D.: 1997, *The River Dragon has come!: Three Gorges dam and the fate of China's Yangtze River and its people*, ME Sharpe.

Ralston, D.: 1987, Mechanics of embankment erosion during overflow, Proc. of the ASCE National Conference in Hydraulic Engineering, pp. 733–738.

Ré, R.: 1946, Etude du lacher instantané d'une retenue d'eau dans un canal par la méthode graphique (Study of the Sudden Water Release from a Reservoir in a Channel by a Graphical Method.), *Jl La Houille Blanche (in French)* **1**(3), 181–187.

Ritter, A.: 1892, Die Fortpflanzung der Wasserwellen (Propagation of waves), *Zeitschrift des Vereines Deutscher Ingenieure* **36**(33), 947–954.

Robinson, K. M. and Hanson, G. J.: 1994, A deterministic headcut advance model, *Transactions of the ASAE* **37**(5), 1437–1443.

Rogers, J. D.: 2007, Background on failure of Teton Dam near Rexburg, Idaho on June 5, 1976, University of Missouri-Rolla, USA. Retrieved from web.umr.edu/~rogersda/teton_dam.

Rosa, L. P. and dos Santos, M. A.: 2000, Certainty & uncertainty in the science of greenhouse gas emissions from hydroelectric reservoirs, Report prepared for the World Commission on Dams.

Rosenblatt, F.: 1958, The perceptron: A probabilistic model for information storage and organization in the brain., *Psychological Review* **65**(6), 386–408.

Rousseeuw, P. J.: 1984, Least median of squares regression, *Journal of the American Statistical Association* **79**(388), 871–880.

Rousseeuw, P. J. and Leroy, A. M.: 2003, *Robust regression and outlier detection*, Wiley-Interscience.

Rowe, W. D.: 1994, Understanding uncertainty, *Risk Analysis* **14**(5), 743–750.

Rozov, A. L.: 2003, Modeling of washout of dams, *Journal of Hydraulic Research* **41**(6), 565–577.

Saltelli, A.: 2004, Global Sensitivity Analysis: An Introduction, pp. 27–43.

Saltelli, A., Chan, K. and Scott, E. M.: 2000, *Sensitivity analysis*, Wiley.

Samuels, P., Klijn, F. and Dijkman, J.: 2005, River flood risk management policies in different parts of the world: Synthesis of special session by NCR, Proc. of the 3rd International Symposium on Flood Defence, Nijmegen, the Netherlands.

Schoklitsch, A.: 1917, Über Dambruchwellen, *Sitzungberichten der Königliche Akademie der Wissenschaften, Vienna, Austria* **126**, 1489–1514.

Schoklitsch, A.: 1926, *Geschiebebewegung in Flüssen und an Stauwerken*, J. Springer.

Schoklitsch, A.: 1934, Der Geschiebetrieb und die Geschiebefracht, Wasserkraft und Wasserwirtschaft, 29, Jahrgang.

Shortreed, J., Dinnie, K. and Belgue, D.: 1995, Risk criteria for public policy, Proc. of the 1st Biennial Conference on "Process Safety and Loss Management in Canada", Institute for Risk Research, University of Waterloo, Waterloo, Ontario, Canada, pp. 131–158.

Simonovic, S. P. and Ahmad, S.: 2005, Computer-based model for flood evacuation emergency planning, *Natural Hazards* **34**(1), 25–51.

Singh, K. P. and Snorrason, A.: 1982, Sensitivity of outflow peaks and flood stages to the selection of dam breach parameters and simulation models, *Technical Report 289*, State Water Survey Division at the University of Illinois, USA.

Singh, K. P. and Snorrason, A.: 1984, Sensitivity of outflow peaks and flood stages to the selection of dam breach parameters and simulation models, *Journal of Hydrology* **68**, 295–310.

Singh, V. P.: 1996, *Dam breach modeling technology*, Kluwer Academic Pub.

Singh, V. P. and Quiroga, C. A.: 1987, A dam-breach erosion model: I. Formulation, *Water Resources Management* **1**(3), 177–197.

Smart, G. M.: 1984, Sediment transport formula for steep channels, *Journal of Hydraulic Engineering* **110**(3), 267–276.

Soares Frazão, S.: 2002, *Dam–break induced flows in complex topographies: Theoretical, numerical and experimental approaches*, PhD thesis, Université Catholique de Louvain, Belgium.

Solomatine, D. P.: 2002, *Applications of data-driven modelling and machine learning in control of water resources*, Computational Intelligence in Control, Idea Group Publishing, pp. 197–217.

Soumis, N., Lucotte, M., Duchemin, a., Canuel, R., Weissenberger, S., Houel, S. and Larose, C.: 2005, *Hydroelectric reservoirs as an anthropogenic sources of greenhouse gases*, John Wiley & Sons, Ohio.

Spinewine, B. and Zech, Y.: 2007, Small-scale laboratory dam-break waves on movable beds, *Journal of Hydraulic Research* **45**(Extra Issue), 73–86.

Stelling, G. S. and Duinmeijer, S. P. A.: 2003, A staggered conservative scheme for every Froude number in rapidly varied shallow water flows, *International Journal for Numerical Methods in Fluids* **43**(12), 1329–1354.

Stelling, G. S., Kernkamp, H. W. J. and Laguzzi, M. M.: 1998, Delft flooding system: a powerful tool for inundation assessment based upon a positive flow simulation, Proc. of 5th International Conference on Hydroinformatics, pp. 449–456.

Stigler, S. M.: 1981, Gauss and the invention of least squares, *The Annals of Statistics* **9**(3), 465–474.

Stoker, J. J.: 1957, Water waves: The mathematical theory with applications, *Pure and Applied Mathematics* **4**.

TAW: 1998, Fundamentals on water defences, *Technical report*, Technical Advisory Committee on Water Defences, the Netherlands.

TAW: 2002, Wave run-up and wave overtopping at dikes, *Technical report*, Technical Advisory Committee on Water Defences, the Netherlands.

Temple, D. M. and Hanson, G. J.: 1994, Headcut development in vegetated earth spillways, *Applied Engineering in Agriculture* **10**(5), 677–682.

Temple, D. and Moore, J.: 1997, Headcut advance prediction for earth spillways, *Transactions of the ASAE* **40**(3), 557–562.

Thoft–Christensen, P. and Baker, M. J.: 1982, *Structural reliability theory and its applications*, Springer.

Tingsanchali, T. and Hoai, H. C.: 1993, Numerical modelling of dam surface erosion due to flow overtopping, Vol. 1, Advances in HydroScience and Engineering, Washington, USA, pp. 883–890.

Tinney, R. E. and Hsu, H. Y.: 1961, Mechanics of washout of an erodible fuse plug, *Journal of the Hydraulics Division* **87**(3), 1–30.

Toro, E. F.: 1999, *Rieman solvers and numerical methods for fluid dynamics: A practical introduction*, Springer-Verlag, Heidelberg, Berlin.

Tremblay, A., Lambert, M. and Gagnon, L.: 2004, Do hydroelectric reservoirs emit greenhouse gases?, *Environmental Management* **33**, 509–517.

Turing, A. M.: 1950, Computing machinery and intelligence, *Mind* **59**(236), 433–460.

U.S. Army Corps of Engineers: 1960, Floods resulting from suddenly breached dams: Conditions of minimum resistance, Report 1(374), U.S. Army Engineers Waterways Experiment Station, Vicksburg, Mississippi.

U.S. Army Corps of Engineers: 1961, Floods resulting from suddenly breached dams: Conditions of high resistance, Report 2(374), U.S. Army Engineers Waterways Experiment Station, Vicksburg, Mississippi.

U.S. Bureau of Reclamation: 1988, *Downstream hazard classification guidelines*, ACER Technical Memorandum No. 11, Assistant Commissioner-Engineering and Research, U.S. Department of the Interior, Denver, Colorado.

van de Ven, G. P.: 2004, *Man-made lowlands: History of water management and land reclamation in the Netherlands*, Matrijs, Utrecht, the Netherlands.

van der Sande, C. J., de Jong, S. M. and de Roo, A. P. J.: 2003, A segmentation and classification approach of IKONOS-2 imagery for land cover mapping to assist flood risk and flood damage assessment, *International Journal of Applied Earth Observation and Geoinformation* 4(3), 217–229.

van der Sluijs, J. P.: 1997, *Anchoring amid uncertainty: On the management of uncertainties in risk assessment of anthropogenic climate change*, PhD thesis, Utrecht University, the Netherlands.

van Rijn, L. C.: 1993, *Principles of sediment transport in rivers, estuaries and coastal seas*, Aqua Publications.

van Schijndel, S.: 2005, The Planning Kit, a decision making tool for the Rhine branches, Proc. of the 3rd International Symposium on Flood Defence, Nijmegen, the Netherlands.

Vaskinn, K. A.: 2003, Stability and failure mechanisms of dams, Proc. of the 3rd IMPACT Workshop, Université Catholique de Louvain, Louvain-la-Neuve, Belgium.

VenW: 2005, Floris study-Full report, Flood Risks and Safety in the Netherlands (Floris) Project, Dutch Ministry of Transport, Public Works and Water Management.

Verheij, H. J.: 2002, Modification breach growth model in HIS–OM, Q3299, *Technical report*, WL | Delft Hydraulics (in Dutch).

Verwey, A.: 2001, Latest developments in floodplain modelling–1D/2D integration, Proc. of 6th Conference on Civil Engineering Hydraulics, Hobart, Australia.

Vesely, W. E. and Rasmuson, D. M.: 1984, Uncertainties in nuclear probabilistic risk analyses., *Risk Analysis* 4(4), 313–322.

Vis, M., Klijn, F., de Bruijn, K. M. and van Buuren, M.: 2003, Resilience strategies for flood risk management in the Netherlands, *International Journal of River Basin Management* 1, 33–40.

Visser, P. J.: 1998, *Breach growth in sand–dikes*, PhD thesis, Delft University of Technology.

Visser, P. J., Smit, M. J. and Snip, D. W.: 1996, Zwin '94 experiment: Meetopstelling en overzicht meetresultaten, Report 4–96, Delft University of Technology, the Netherlands.

Vrijling, J. K.: 2001, Probabilistic design of water defense systems in the Netherlands, *Reliability Engineering and System Safety* **74**(3), 337–344.

Vrouwenvelder, A. C. W. M.: 1999, Theoretical manual of PC-Ring, Part C: Calculation methods (in Dutch), TNO-report 98-CON-R1204, Delft, the Netherlands.

Vrouwenvelder, A. C. W. M., Steenbergen, H. M. G. M. and Slijkhuis, K. A. H.: 2001, Theoretical manual of PC-Ring, Part B: Statistical models (in Dutch), TNO-report 98-CON-R1431, Delft, the Netherlands.

Wahl, T. L.: 1998, Prediction of embankment dam breach parameters: A literature review and needs assessment, U.S. Department of the Interior, Bureau of Reclamation, Dam Safety Office.

Wahl, T. L.: 2004, Uncertainty of predictions of embankment dam breach parameters, *Journal of Hydraulic Engineering* **130**.

Wentz, F. J., Ricciardulli, L., Hilburn, K. and Mears, C.: 2007, How much more rain will global warming bring?, *Science* **317**(5835), 233–235.

Witten, I. H. and Frank, E.: 2000, *Data mining: Practical machine learning tools and techniques with Java implementations*, Morgan Kaufmann.

Wolters, H. A., Platteeuw, M. and Schoor, M. M.: 2001, Guidelines for rehabilitation and management of floodplains, NCR publication.

World Commission on Dams: 2000, *Dams and development: A new framework for decision-making*, Earthscan.

Wu, W., Wang, S. S. Y., Jia, Y. F. and Robinson, K. M.: 1999, Numerical simulation of two-dimensional headcut migration, ASCE, Proc. of the International Water Resources Engineering Conference, Seattle, WA, USA.

Yalin, M. S.: 1972, *Mechanics of sediment transport*, Pergamon Press.

Zagonjolli, M., Goossens, H. and Mynett, A. E.: 2006, A numerical-constraint based approach for flood mitigation in dike ring areas, Proc. of 7th International Conference on Hydroinformatics, Nice, FRANCE.

Zagonjolli, M. and Mynett, A.: 2006a, Dam breaching uncertainty and its effect in downstream areas, Proc. of the 7th International Conference on Hydroscience and Engineering, Philadelphia, USA, Michael Piasecki and College of Engineering, Drexel University.

Zagonjolli, M. and Mynett, A. E.: 2005a, Dam breach analysis: A comparison between physical, empirical and data mining models, Proc. of the 29th IAHR Congress Seoul, South Korea, p. 753.

Zagonjolli, M. and Mynett, A. E.: 2005b, Data mining techniques in dam breach modelling, *Submitted to the IAHR Journal of Hydraulic Research* .

Zagonjolli, M. and Mynett, A. E.: 2006b, Failure analysis of a high earthfill dam: Scenario development using numerical simulation and GIS, Proc. of 7th International Conference on Hydroinformatics, Nice, FRANCE.

Zagonjolli, M., Mynett, A. E. and Verwey, A.: 2005, Dam break modelling analysis of Bovilla Dam near Tirana, Albania, Proc. of the 3rd International Symposium on Flood Defence, Nijmegen, the Netherlands.

Zhu, Y., Visser, P. J. and Vrijling, J. K.: 2006, Laboratory observations of embankment breaching, Proc. of the 7th International Conference on Hydroscience and Engineering, Philadelphia, USA, Michael Piasecki and College of Engineering, Drexel University.

List of Symbols

Roman Letters

Symbol	Description	SI Units
a_{bl}	Thickness of the bed layer	[m]
B_{avg}	Ultimate average width of the breach channel	[m]
B_{bot}	Bottom width of the breach channel	[m]
B_t	Top width of the breach channel	[m]
C	Weir discharge coefficient	[-]
C_s	Submergence coefficient on the weir	[-]
d	Soil grain diameter	[mm]
D	Height of the breach channel	[m]
DF	Dam factor	[m^4]
d_{50}, d_{65}	Representative grain size diameter	[mm]
Fr	Froude number	[-]
g	Gravitational acceleration	[m^2/s]
h	Flow depth	[m]
H	Water depth above the breach bottom	[m]
h_b	Vertical extent of the breach channel	[m]
h_d	Dam height or dike height	[m]
h_w	Depth of water in the reservoir triggering the failure	[m]
k and a	Correlation coefficients	[-]
I_1 and I_2	Integral quantities depending upon the flow depth	[-]
L	Length of the breach channel	[m]
M_s	Mass of soil eroded during breaching	[kg/s]
q	Discharge per unit width	[m^2/s]
Q_p	Peak discharge	[m^3/s]
t_d	Duration of failure	[hr]
u	Water velocity	[m/s]
V_{out}	Volume of water outflow from the reservoir	[m^3]
V_w	Volume of water in the reservoir triggering the failure	[m^3]
Y	Elevation of breach bottom	[m]
z	Breach side slope factor	[-]

Greek letters

Symbol	Description	SI Units
β, δ	Angle that the breach side creates with vertical	[°]
γ	Unit weight of water	[N/m^3]

τ	Tractive stress of the flowing water	$[N/m^2]$
τ_c	Critical tractive stress for erodible material	$[N/m^2]$
$(\tau_0 - \tau_c)$	Excess shear stress	$[N/m^2]$

Mathematical symbols and operators

Symbol	*Description*
$\equiv$	Equivalent to (or defined to be)
$\underset{x}{\mathrm{argmax}}\, f(x)$	Value of x that leads to the maximum value of f(x)

List of Abbreviations

AI	Artificial Intelligence
ANNs	Artificial Neural Networks
BREADA	BReaching of the EArthfill DAm
Floris	FLOod RIsks and Safety
GHGs	GreenHouse Gases
GIS	Geographical Information System
IBL	Instance Based Learning
IPCC	Intergovernmental Panel on Climate Change
LMS	Least Median of Squares
LS	Least Squares
LWL	Locally Weighted Learning
MLP	Multi Layer Perceptron
NPDP	National Performance of Dams Program in USA
PAR	People At Risk
RL	Reinforcement Learning
RMSE	Root Mean Square Error
SMDBRK	National Weather Service Simplified Dam Break Model
TAW	Technical Advisory Committee on Water Defences in the Netherlands
USBR	U.S. Bureau of Reclamation
WCD	World Commission on Dams
WEKA	Waikato Environment for Knowledge Analysis

Curriculum Vitae

Migena Zagonjolli was born in 1975 in Tirana, Albania. She studied hydraulic engineering at the Polytechnic University of Tirana, Albania, where she graduated in 1999. From then until 2001 she worked at the Department of Hydraulic Engineering as an assistant lecturer in Hydraulics and Water Resources Management. During this period she visited the University of Limerick, Republic of Ireland, as a guest researcher.

From 2001 to 2003 she studied at UNESCO–IHE in Delft, the Netherlands, where she obtained her MSc degree in Hydroinformatics on a research topic related to dam break modelling. In 2004 she started her PhD research within the collaboration framework between WL | Delft Hydraulics, UNESCO–IHE and Delft University of Technology. She addressed a range of research topics, including data mining and analysis, computational modelling of dam and dike breaching, sediment transport, and flood risk assessment. One of the novel aspects of her research is the design of a numerical–constraint based model for flood propagation and risk mitigation.

Her experience includes software development related to dam/dike breaching, flood wave propagation, constraint based modelling, uncertainty analysis and flood mitigation. She has published several papers in international journals and conference proceedings. Starting 1st October 2007 she will be employed as a researcher/advisor at the Department of Inland Water System at WL | Delft Hydraulics.

Printed and bound by CPI Group (UK) Ltd, Croydon, CR0 4YY

22/10/2024

01777530-0017